Natural Disaster Analysis after Hurricane Katrina

Risk Assessment, Economic Impacts and Social Implications

Edited by

Harry W. Richardson, Peter Gordon and
James E. Moore II

University of Southern California, USA

Edward Elgar
Cheltenham, UK • Northampton, MA, USA

Published by
Edward Elgar Publishing Limited
The Lypiatts
15 Lansdown Road
Cheltenham
Glos GL50 2JA
UK

Edward Elgar Publishing, Inc.
William Pratt House
9 Dewey Court
Northampton
Massachusetts 01060
USA

Paperback edition 2009

A catalogue record for this book
is available from the British Library

Library of Congress Control Number: 2008927953

ISBN 978 1 84720 357 1 (cased)
ISBN 978 1 84844 776 9 (paperback)

Printed and bound by MPG Books Group, UK

Contents

List of contributors   vii
Preface and acknowledgments   ix

1   Introduction   1
    Harry W. Richardson, Peter Gordon and James E. Moore II
2   Comprehensive disaster insurance: will it help in a post-
    Katrina world?   8
    Howard C. Kunreuther and Erwann O. Michel-Kerjan
3   A decision analysis of options to rebuild the New Orleans
    flood control system   34
    Carl Southwell and Detlof von Winterfeldt
4   Hurricane Katrina: lessons learned   53
    Jiin-Jen Lee and Bennington Willardson
5   Katrina vs. 9/11: how should we optimally protect against
    both?   71
    Jun Zhuang and Vicki M. Bier
6   Worst-case thinking and official failure in Katrina   84
    Lee Clarke
7   Risk, preparation, evacuation and rescue   93
    Edd Hauser, Sherry M. Elmes and Nicholas J. Swartz
8   Not Katrina: the Thames Barrier decision   120
    Chang-Hee Christine Bae and Harry W. Richardson
9   Is New Orleans ready to celebrate after Katrina? Evidence
    from Mardi Gras and the tourism industry   134
    Kathleen Deloughery
10 Estimating the state-by-state economic impacts of Hurricane
    Katrina   147
    Jiyoung Park, Peter Gordon, James E. Moore II, Harry
    W. Richardson, Soojung Kim and Yunkyung Kim
11 Regional economic impacts of natural and man-made
    hazards: disrupting utility lifeline services to households   187
    Adam Rose and Gbadebo Oladosu
12 Adjusting to natural disasters   208
    V. Kerry Smith, Jared C. Carbone, Jaren C. Pope, Daniel G.
    Hallstrom and Michael E. Darden

13 Katrina: a Third World catastrophe?   230
    Edward J. Clay
14 Hurricane Katrina and housing: devastation, possibilities
    and prospects   253
    Raphael W. Bostic and Danielle Molaison
15 Unnatural disaster: social impacts and policy choices after
    Katrina   279
    John R. Logan

Index   299

Contributors

Chang-Hee Christine Bae, Department of Urban Design and Planning, University of Washington, Seattle

Vicki M. Bier, University of Wisconsin-Madison

Raphael W. Bostic, School of Policy, Planning and Development, University of Southern California

Jared C. Carbone, Williams College

Lee Clarke, Rutgers University, Brunswick, NJ

Edward J. Clay, Overseas Development Institute, London

Michael E. Darden, University of North Carolina at Chapel Hill

Kathleen Deloughery, Ohio State University

Sherry M. Elmes, University of North Carolina at Charlotte

Peter Gordon, School of Policy, Planning and Development, University of Southern California

Daniel G. Hallstrom, North Carolina State University

Edd Hauser, University of North Carolina at Charlotte

Soojung Kim, Millken Institute, Santa Monica, CA

Yunkyung Kim, School of Policy, Planning and Development, University of Southern California

Howard C. Kunreuther, Risk Management and Decision Processes Center, Wharton School, University of Pennsylvania

Jiin-Jen Lee, Viterbi School of Engineering, University of Southern California

John R. Logan, Department of Sociology, Brown University, Rhode Island

Erwann O. Michel-Kerjan, Risk Management and Decision Processes Center, Wharton School, University of Pennsylvania

Danielle Molaison, School of Policy, Planning and Development, University of Southern California

James E. Moore II, Department of Industrial and Systems Engineering, Viterbi School of Engineering, University of Southern California

Gbadebo Oladosu, Oak Ridge National Laboratory, Oak Ridge, TN

Jiyoung Park, Post-Doctoral Fellow, CREATE, University of Southern California

Jaren C. Pope, North Carolina State University

Harry W. Richardson, The James Irvine Chair of Urban and Regional Planning, School of Policy, Planning and Development, University of Southern California

Adam Rose, Visiting Professor, School of Policy, Planning and Development, University of Southern California

V. Kerry Smith, Department of Economics, University of Arizona

Carl Southwell, School of Policy, Planning and Development, University of Southern California

Nicholas J. Swartz, University of North Carolina at Charlotte

Bennington Willardson, Viterbi School of Engineering, University of Southern California

Detlof von Winterfeldt, Director, CREATE, Viterbi School of Engineering, University of Southern California

Jun Zhuang, University of Wisconsin-Madison

Preface and acknowledgments

Although it will be obvious to many readers, the primary reason why this book that deals with a natural disaster (Hurricane Katrina) comes out of a research center dealing with terrorism (CREATE – the Center for Risk and Economic Analysis of Terrorism Events at the University of Southern California) is because FEMA (the Federal Emergency Management Agency) is housed within the United States Department of Homeland Security (DHS), and dealing with the consequences of Katrina has been a major concern for the DHS. However, a second reason is that natural disaster and man-made disaster (that is, terrorism) research has substantial overlap, and many researchers (including some in this book) work in both areas. We are grateful to CREATE for financial support and to its Director Detlof von Winterfeldt for his continued support for our activities.

The research in this book was supported by the Department of Homeland Security through the Office of Naval Research under the grant number N00014-05-1-0630. However any opinions, findings and conclusions or recommendations in the book are those of the authors and do not necessarily reflect views of the US Department of Homeland Security.

The chapter 'Adjusting to natural disasters' originally appeared in the *Journal of Risk and Uncertainty*, 33, 37–54 (Smith et al., 2006) and is used here with kind permission from Springer Science and Business Media.

A more technical version of Chapter 5, 'Katrina vs. 9/11: how should we optimally protect against both?', appeared in *Operations Research*, **55** (5), 976–91.

1. Introduction

Harry W. Richardson, Peter Gordon and
James E. Moore II

This book explores some of the issues arising from the Hurricane Katrina disaster. Scholars from many fields such as decision analysis, risk management, economics, engineering, transportation, urban planning and sociology explore some of the more important policy issues resulting from Katrina such as insurance, flood control and rebuilding the levees, housing, tourism, evacuation and relocation, utility lifelines recovery and resilience, the racial implications of the disruption of life in New Orleans, and interregional economic impacts of the disaster. The focus is less on what happened in the past than on how to deal with future risks, not only in New Orleans but in other locations threatened by disaster.

Of course, the travails of New Orleans in the wake of Katrina are far from over. There is also a mound of research already in the pipeline on Katrina (some of it already published), and it is impossible for a single book such as this to cover the full range of topics comprehensively, especially when authors from so many disciplines are involved.

There are two chapters that deal with levee rebuilding (Chapters 3 and 4) while a third (Chapter 8) addresses the issue of flood protection in a very different context (the River Thames in London, England). However, they do not fully answer the question of how to trade off between high construction costs and category 5 hurricane protection.

Another major issue barely mentioned in the book is whether rebuilding New Orleans at all is a wise idea. There is a case for building a new New Orleans on higher ground further north in the state. Despite all the policy failures since Katrina, this is probably a political non-starter. So, an intermediate position is that New Orleans will end up much smaller than before. Many of the former residents will not come back, and there is little point (given the risks) in attracting new residents.

Kunreuther and Michel-Kerjan (Chapter 2) examine the role that insurance can play as a policy tool for reducing losses from future natural disasters while at the same time providing funds for recovery. After examining the decision processes of three categories of interested parties who will be at the centerpiece of such an insurance program – residents in hazard-prone areas, insurers and reinsurers, and the federal government – they provide a rationale for comprehensive disaster insurance with rates based on risk as an integral part of a hazard management program. They illustrate how such a program would function in the context of a case study of the Florida market. To reduce future losses there is a need for creative private–public partnerships to complement insurance through economic incentives and well-enforced regulations and standards (for example, building codes). The chapter also explores the issues as to whether coverage should be voluntary or mandatory and what types of special arrangements should be given to low-income families in high-hazard areas.

Southwell and von Winterfeldt (Chapter 3) remind us that when the Corps of Engineers analyzed several alternatives for improving the levee systems around New Orleans in the 1970s and 1980s, they made several rather optimistic assumptions about the frequency and impacts of hurricanes. They also ignored breaches of the levees as a failure mode. As a result, they concluded that a protection level for a category 3 hurricane would be sufficient. This chapter builds upon an earlier decision analysis framework for improving how to rebuild the levee system after Katrina. It reports on a formal, fully parameterized risk and cost–benefit analysis model using influence diagrams. Some preliminary conclusions are presented.

Lee and Willardson (Chapter 4) point out that the Katrina disaster was not merely the result of the overtopping and breaches of the levees but also the product of electrical power failure and the abandonment of the pumping stations. For example, as explained in Chapter 8, the Thames Barrier has its own electric generators and sells surplus power to power stations when their system is not in active use. The flood protection system in the New Orleans area is quite complex, but it is only as strong as its weakest link. In case of failure, there is a need for more effective recovery mechanisms. Scarce resources have dictated repairs below the category 5 level, so the risks of a recurrence have not been fully addressed.

Bier and Zhuang (Chapter 5) describe the results of a model that applies game theory to identify equilibrium strategies for both attacker and defender in a fully endogenous model of resource allocation for countering both terrorism and natural disasters. The key novel features of this model include balancing protection from terrorism and natural disasters, and describing the attacker choice by a continuous level of effort rather than a discrete choice (for example, attack or not). Interestingly, in a sequential game, increased defensive investment can lead an attacker either to increase the level of effort (to help compensate for the reduced probability of damage from an attack), or to decrease the level of effort (because attacking has become less profitable). This can either reduce or increase the effectiveness of investments in protection from intentional attack, and can therefore affect the relative desirability of investing in protection against natural disasters.

Clarke (Chapter 6) argues that considering worst-case scenarios helps to design more effective policy. Sharing the blame with everybody was very popular with federal officials (that is, those who had the mandate and enough power to have ameliorated the hurricane's damage). The problem with this position is that it distributes responsibility evenly. That is a mistake because if power, among officials and among organizations, is distributed unevenly then responsibility should be too.

One of the biggest official failures of 9/11 was one of imagination. That failure led the US government to be caught off guard by what to many others seemed an obvious risk. But what is a worst-case imagination? It is one that emphasizes possibilities over probabilities; it emphasizes the consequences of the likelihood that courses of events will occur. Probabilistic thinking has come to be equated with rationality itself. But this is a mistake. Possibilistic thinking can be usefully employed to counterbalance probabilism. There is some relationship between possibilism and counterfactual thinking. Explore a worst-case counterfactual world, and prepare for it.

Hauser et al. (Chapter 7) take account of the fact that recent, current and future threats to the American homeland, whether by natural elements or by human-induced threats or actions, represent unique and complex strategic and tactical planning challenges that must be well understood by local, state and federal emergency responders, including the military. Considerable attention is currently being given to research and development into modified trip generation, distribution, route and modal assignment transportation models, with hurricane evacuation modeling the current focus.

To be valid, decision support tools must account for the dynamic and adaptive interactions of the threat and the populations at risk. Individual-based computational simulation modeling holds considerable promise to provide a framework for simulating and analyzing the non-linear macroscopic impacts of the numerous microscopic interactions within the population at risk (in both known and unknown scenarios). The potential for loss of life, serious injury, and the physical, economic and psychological aspects of catastrophes need to be estimated *ex ante* in order to make mitigating real-time decisions.

The research aims to understand better why decisions are made about whether to evacuate or not to evacuate. About 400 residents or former residents of New Orleans and Orleans Parish were interviewed about their decisions. Geographic locations, socio-economic status and mobility characteristics were all examined. The results of the study can be used to develop a risk profile to help local authorities decide whether to designate and enforce a mandatory evacuation, and the optimal time to do it.

Bae and Richardson (Chapter 8) take a more detached and more distant look from New Orleans to examine a case where hitherto flood protection has worked: the Thames Barrier in London. Many world cities have been prone to floods for centuries. Although none of them can prevent floods, some have been much more successful in dealing with floods and protecting land than others. What makes the difference? Acknowledgment of the costs and the willingness to bear them plus leadership. The project was spear-headed by a Scientific Advisor to the government, Sir Herman Bondi, in 1966 although the Barrier was not opened until 1983. For the future, there is a new proposal to block the 10 miles width of the River Thames reaching from Kent to Essex, and to build a road above the dam. This is a very ambitious project to protect Londoners from flooding related to an anticipated 2–3 ft sea level rise, possibly associated with greenhouse gas emissions.

Deloughery (Chapter 9) examines one of the most important sectors in New Orleans, tourism, and its prime event, Mardi Gras. Many viewed the 2006 Mardi Gras celebration as a way to gauge the early economic rebirth of New Orleans. The chapter looks at data on Mardi Gras since 1990 to determine the effect Hurricane Katrina had on the 2006 celebration. First, hotel data is examined. With the damage to downtown New Orleans, which hotels were able to open? And, of those, how many were able to operate close to capacity? Taking the capacity constraints into consideration, Deloughery looks at whether people were still staying in the city, or if the crowd had moved to suburban areas. The main indicator of whether Mardi Gras was as big as ever, and that New Orleans is on the road to recovery, is the amount of money people spent while there. In order to look at the full spectrum of the New Orleans economy, revenue is considered from many different sectors. Bar, restaurant, casino and transportation (cabs and trolleys) revenue are examined for the last decade. However, the fact that many establishments had not reopened should be taken into consideration when drawing conclusions from these data.

One of the most celebrated features of Mardi Gras is the parades. These parades are run primarily by those living in and around New Orleans. Looking at the number of parades, and the number of floats in each parade, yields information about the number of Louisiana natives coming to Mardi Gras. This facilitates a breakdown of the effect between natives and tourists on the economy of New Orleans during Mardi Gras. The Mardi Gras celebration is usually a two-week event, becoming more crowded as Fat Tuesday draws near. Determining the length of time people are staying in New Orleans, and when the influx of people starts to arrive, yielded more information about whether the 2006 Mardi Gras celebration had been scaled down. Additionally, these results can be compared with other New Orleans events popular with tourists, such as the Jazz Festival. This comparison sheds light on whether the Mardi Gras is a true representation of the economic growth of New Orleans, or whether it is an outlier.

Park et al. (Chapter 10) apply the National Interstate Economic Model (NIEMO), the first operational multi-state input–output model of the US, to estimate the state-by-state economic impacts of Hurricane Katrina. Whereas the effects of Katrina on US gross domestic product (GDP) have been minor, this sort of aggregation masks important sub-national differences. Estimates of direct losses of tourism, port services, and oil and gas refining drive the model. The results are for 50 states, 47 industrial sectors and each of the three major sectors.

Rose and Oladosu (Chapter 11) analyze resilience. The term refers to the ability to mute potential economic losses from disasters through various types of inherent and adaptive coping mechanisms. These take place at three levels: individual business (for example, conservation, input substitution, relocation), market (for example, resource allocating ability of prices), and regional economy (for example, import substitution). Several studies indicate that resilience in the case of ordinary disasters is quite high (see, for example, Tierney, 1997; Rose and Lim, 2002; Rose and Liao, 2005, all referenced at the end of Chapter 11). The studies to date, however have been limited to the effects of a disaster targeting one sector (for example, terrorist attacks on electricity generators or transmission lines) or in large cities with relatively small pockets of damage (for example, the Northridge earthquake in Los Angeles, and 11 September 2001 in New York City).

The situation in the aftermath of Hurricane Katrina, however, is far different from these other events. The physical damage is so widespread and recovery is so slow that these factors may have undercut or overwhelmed resilience. For example, inventories have been depleted, substitution options limited, permanent loss of customers and suppliers limit production rescheduling, some markets may become so thin or disrupted that they can no longer be depended on to provide the proper signals for efficient resource allocation, and disrupted transport networks limits imports flowing into the region.

The purpose of the research is to use data collected from the Hurricane Katrina experience to analyze the effectiveness of resilience in the context of a true catastrophe. Implications are drawn about the effects of disaster magnitude and duration on resilience. Another focus is on how resilience can be enhanced or safeguarded so that it will be more effective against future catastrophes.

People adjust to the risks presented by natural disasters in a number of ways: they can move out of harm's way, they can self-protect, or they can insure. Smith et al. (Chapter 12) examine Hurricane Andrew, the largest US natural disaster prior to Katrina, to evaluate how people and housing markets respond to a large disaster. The analysis combines a unique *ex post* database on the storm's damage along with information from the 1990 and 2000 Censuses in Dade County, Florida where the storm hit. The results suggest that the economic capacity of households to adjust explains most of the differences in demographic groups' patterns of adjustment to the hurricane damage. Low-income households respond primarily by moving into low-rent housing in areas that experienced heavy damage. Middle-income households move away to avoid risk, and the wealthy, for whom insurance and self-protection are most affordable, appear to remain. This pattern of adjustment with respect to income is roughly mean-neutral, so an analysis based on measures of central tendency such as median income would miss these important adjustments.

Clay (Chapter 13) argues that Hurricane Katrina could prove to be a defining event in the way the global community thinks about natural disasters. The conventional wisdom has been that, with development, disaster risk – the combination of hazard probabilities and vulnerabilities – declines. Richer countries invest in preventive measures, engage in risk transfer and evolve more effective public and private responses to natural hazards. Hurricane Katrina dramatically calls in question such assumptions.

We already know that there was institutional failure at federal, state and local government levels. Some of the problems that have been exposed involve at least similarities to disasters in poorer, so-called developing countries. First, the professional risk assessments pointing to the serious possibility of a catastrophic event did not result in commensurate disaster prevention measures with regard to either regulation of human habitat and business activity or investment levels of storm and flood protection. Second, the public response revealed problems similar to those found in the manuals on disaster risk reduction for developing countries. Third, analyses for developing countries typically focus on the links among vulnerability, poverty, social exclusion and marginality. An interesting question is whether we are dealing with problems of a general nature, inherent in the way both public institutions and the private sector manage risk and respond to catastrophic events. Are some developing countries, with much more limited resources and that are exposed to similar categories of hazard, more effective in at least some aspects of disaster prevention or response? Are higher levels of economic development likely to be associated with structural changes in the economy and society that could increase some of the risks associated with extreme, highly improbable natural events? Clay considers and contrasts evidence from developing counties exposed to extreme riverine flood and coastal hazards with Hurricane Katrina and the 2002 Elbe floods in Central Europe.

Bostic and Molaison (Chapter 14) examine Katrina's impact on the New Orleans housing sector and assesses the various alternatives for reconstruction of the city. The chapter begins by providing a broad inventory of the damage wrought by the storm, and focuses on the spatial distribution of the damage and the resultant implications for reconstruction in terms of city functionality and cost. The analysis then turns to the various plans for reconstruction and evaluates them in light of the damage inventory and its implications. It concludes by highlighting general lessons learned from the Katrina experience that can inform policy-makers and emergency responders to enhance preparation for future disasters to minimize housing-related damage and dislocation.

Logan (Chapter 15) reports the preliminary results of a study of the impacts of Hurricane Katrina on New Orleans, focusing mainly on the future redevelopment of the city. The key questions are: whose neighborhoods were destroyed, and whose neighborhoods will be rebuilt? Analyses of the pattern of development of New Orleans in the second half of the twentieth century show how race and class became closely connected to elevation in a city where there was always a risk of water damage.

Research on the post-Katrina situation provides information on the pattern of dispersion and return of residents, the impacts of early policy decisions on recovery, and the mobilization of neighborhood constituencies to participate in the political process.

The post-Katrina saga will continue for some years to come. It is unknown what the eventual stable population will be. Resolution of the housing stock problem will take time, and again with an indeterminate outcome. The Louisiana Recovery Authority still has much work to do. The funding stream from the federal government is intermittent and uncertain. The strength of the repaired levees is problematic. There is a powerful case for the drastic reform of the Federal Emergency Management Agency (FEMA). There are many lessons still to be learned. Our hope is that this book makes a contribution to the debates.

2. Comprehensive disaster insurance: will it help in a post-Katrina world?

Howard C. Kunreuther and
Erwann O. Michel-Kerjan

INTRODUCTION

The White House National Strategy defines homeland security as 'the concerted effort to prevent attacks, reduce America's vulnerability to terrorism, and minimize the damage *and* recover from attacks that do occur' (White House, 2002 and 2007). To succeed, homeland security must be a national and comprehensive effort. Moreover, that definition must apply to technological and natural disasters as well. Indeed, while important efforts have been undertaken since 9/11 by all levels of government and private sector actors to prevent a new terrorist attack, such a focus on terrorism threat might have diverted resources away from the more common types of disasters; that is, natural catastrophes (DHS, 2006).

This chapter discusses how best to finance the economic consequences of large-scale natural disasters by focusing on the role that the insurance infrastructure can play in conjunction with other activities that all levels of government will play as well.

Whereas protecting residential and commercial construction and critical infrastructure services (transportation, telecommunications, electricity and water distribution, and so on) in risky areas may limit the occurrence and/or the impacts of major catastrophes, we know that major natural disasters will still occur. In these situations one must adequately assure emergency measures and rapidly restore critical services. The question as to who will provide financial protection to victims (residents and commercial enterprises) will take center stage. The insurance infrastructure will then play a critical role (Auerswald et al., 2006).

The United States has extensive experience with natural catastrophes. But the recent hurricanes that occurred during the 2004 and 2005 seasons have changed the landscape forever. Indeed, they have demonstrated major failures in the country's preparedness when it comes to dealing with large-scale natural disasters: lack of emergency preparedness capacity, lack of adequate loss reduction measures, and many uninsured people requiring historical levels of federal aid (over $110 billion). Katrina killed 1300 people and forced 1.5 million people to evacuate the affected area – a record for the United States. In addition to property damage, the ripple effect to interruption of economic activities for weeks if not months led to estimates of economic damages in the range of $150 billion to $200 billion.

The increasing losses from recent hurricanes in the United States have forced some major insurers to re-examine their ability to provide financial protection against mega natural disasters – so-called 'super-cats' where insured losses are higher than $10 billion. Some insurers have stopped providing coverage against certain types of risk in certain locations because they are concerned about the impact that another large disaster will have on their assets. Others are restricting the amount of protection they are willing to provide in hazard-prone areas. This raises several critical points that the present chapter addresses.

The first section discusses the radical change in recent loss trends with respect to natural disasters. If we consider the last 37 years (1970–2007), 18 of the 20 most costly disasters for the insurance industry occurred after 1990 and ten of these were after 2000 (nine in the United States alone). Insurance claim payments for natural disasters in the US were twice as high in 2005 as they were in 2004, the previous record holder. The next section provides an overview of different programs in place in the US to deal with disaster insurance. We then propose two principles on which disaster insurance should be based if one wants to avoid a repetition of a Katrina-type disaster: (1) insurance premiums should be based on risk, and (2) protecting low-income residents in hazard-prone areas with insurance. The following section delineates the opportunities and challenges of a risk-based all-hazards comprehensive disaster insurance program.[1] The final section summarizes the key points of the chapter and suggests directions for future research.

A TOTALLY NEW DIMENSION IN INSURED CATASTROPHES

The past few years have seen the emergence of a totally new dimension of natural disasters in many countries. While one used to refer to such hazards as low-probability, high-consequence events, these 'extreme events' seem to be more frequent and more severe now than would have been predicted from past data.

Figure 2.1 depicts the upward trend in worldwide *insured* losses from catastrophes between 1970 and 2007 (in 2007 indexed prices).[2] Between 1970

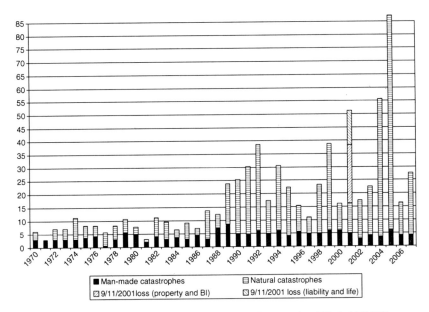

Notes: 9/11: all lines, including property and business interruption (BI); in US$ billion indexed to 2007.

Sources: Wharton Risk Center with data from Swiss Re and Insurance Information Institute.

Figure 2.1   Worldwide evolution of catastrophe insured losses, 1970–2007

and the mid-1980s, annual insured losses from natural disasters (including forest fires) were in the $3 to $4 billion range. The insured losses from Hurricane Hugo, which made landfall in Charleston, South Carolina on 22 September 1989, exceeded $4 billion (1989 prices). It was the first natural disaster to inflict more than one billion dollars of insured losses in the US. There was a radical increase in insured losses in the early 1990s, with Hurricane Andrew in Florida ($23.7 billion in 2007 dollars) and the Northridge earthquake in California ($19.6 billion in 2007 dollars).

Between 1969 and 1998 nearly 650 US insurers, mostly small ones, became insolvent, 50 as a result of natural catastrophes, including nine insurance companies that failed because of Hurricane Andrew in 1992. Moreover, even large insurers were severely impacted by that disaster. For example, the Florida branch of State Farm Fire and Casualty (the largest homeowner insurer in the US) suffered a $4 billion loss (1992 prices) and was rescued by its parent company. The Florida branch of Allstate, the other major player in that market, paid about $1.9 billion in claims from Hurricane

Andrew and was also rescued by its parent company. To put this figure in perspective, the $1.9 billion loss was $500 million more than the Florida operations of Allstate had earned in profits from all types of insurance over the 53 years it had been in business there (IPCC, 2001).

Nature of Insured Losses

Following Hurricane Andrew insurers began to rethink the concept of insurability of a risk. Many utilized catastrophic models to estimate the likelihood and consequences from specific hazards that might cause damage in specific locations (Grossi and Kunreuther, 2005). Since 1990, insurers have continuously improved the way they underwrite catastrophic risks: for example, no insurance company was declared bankrupt as a result of 11 September 2001.

Extreme events have continued to inflict major insured losses from natural disasters. A new record was reached in 2004 with total financial losses in the world of $120 billion, $49 billion of which was covered by insurance (Swiss Re, 2005). This upward trend continued in 2005. Hurricane Katrina alone cost insurers and reinsurers $46.3 billion, and total losses paid by private insurers due to major natural catastrophes were $87 billion in 2005.[3] According to Swiss Re, worldwide major catastrophes in 2005 inflicted $230 billion in economic damage, $83 billion of which was covered by insurance. Despite this historical scale of losses, only one insurance group (Poe Inc.) became insolvent as a result of the 2005 hurricane season, demonstrating the resilience of the insurance industry[4] (Michel-Kerjan, 2006).

At an event-based level, Table 2.1 shows the 20 most costly catastrophes for the insurance sector since 1970 in 2007 dollars. Several observations are important here:

- First, all of these 20 events except for the terrorist attacks on 11 September 2001 were natural disasters (including forest fires).[5] Among the top 19 natural disasters that occurred since 1970, more than 80 percent were weather-related events: hurricanes and typhoons, storms and floods.[6]
- Second, 18 of the 20 most costly events occurred since 1990 (in constant prices). Half of these 20 disasters occurred between 2001 and 2005 and all but one in the US. This new reality obliges insurers and reinsurers to pay much more attention to the catastrophic potential of natural disasters and other extreme events.
- Third, economic losses (insured and non-insured) are largely concentrated in industrialized markets. The evolution of losses from natural disasters in different regions of the world indicates that between 1980

Table 2.1   The 20 most costly insured losses in the world, 1970–2007

U.S.$ billion (indexed to 2007)	Event	Victims (Dead or missing)	Year	Area of primary damage
46.3	Hurricane Katrina	1826	2005	USA, Gulf of Mexico etc.
35.5	9/11 Attacks	3025	2001	USA
23.7	Hurricane Andrew	43	1992	USA, Bahamas
19.6	Northridge Quake	61	1994	USA
14.1	Hurricane Ivan	124	2004	USA, Caribbean et al.
13.3	Hurricane Wilma	35	2005	USA, Gulf of Mexico et al.
10.7	Hurricane Rita	34	2005	USA, Gulf of Mexico et al.
8.8	Hurricane Charley	24	2004	USA, Caribbean et al.
8.6	Typhoon Mireille	51	1991	Japan
7.6	Hurricane Hugo	71	1989	Puerto Rico, USA et al.
7.4	Winterstorm Daria	95	1990	France, UK et al.
7.2	Winterstorm Lothar	110	1999	France, Switzerland et al.
6.1	Winterstorm Kyrill	54	2007	Germany, UK, NL, France et al.
5.7	Storms and floods	22	1987	France, UK et al.
5.6	Hurricane Frances	38	2004	USA, Bahamas
5.0	Winterstorm Vivian	64	1990	Western/Central Europe
5.0	Typhoon Bart	26	1999	Japan
4.5	Hurricane Georges	600	1998	USA, Caribbean
4.2	Tropical Storm Alison	41	2001	USA
4.2	Hurricane Jeanne	3034	2004	USA, Caribbean et al.

Note: This table excludes payments for flood by the National Flood Insurance Program in the US; for example, $17.3 billion in 2005 as a result of Hurricanes Katrina and Rita.

Sources: Wharton Risk Center with data from Swiss Re and Insurance Information Institute.
   Swiss Re, *sigma* No 1/2008 (Natural Catastrophes and Man-Made Disasters in 2007: High Losses in Europe). For more information, see www.swissre.com/sigma. All *sigma* studies are copyright protected, as described on p. 2 of each *sigma* issue.

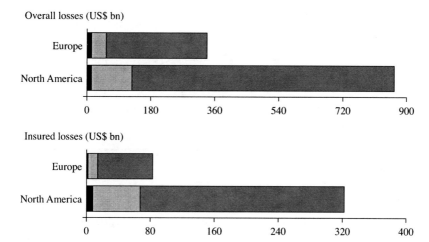

Source: Munich Re (2006).

Figure 2.2 Natural catastrophes 1980–2005: a comparison between Europe and North America

and 2005, North America has accounted for four times more insured losses due to catastrophes than Europe, the second peak zone in the world ($320 billion versus $80 billion for Europe) (see Figure 2.2). The top 20 most devastating events for the insurance industry have inflicted a total of insured losses of $243.1 billion as shown in Table 2.1. The US alone accounts for approximately 80 percent of this total.

Impact on Combined Ratio

The increase in insured losses in recent years needs to be put into perspective by examining the evolution of premiums collected by insurers to cover these risks. As with every other economic sector, insurance has its own vocabulary. An important measure is the 'combined ratio,' defined as the ratio of 'incurred losses (including loss adjustment expenses) plus other expenses' over 'earned premiums.' The combined ratio can be applied to a specific insurer or the entire insurance industry. A combined ratio of 100 indicates that the loss plus expenses equals the premiums – an underwriting-neutral operation. If the ratio is below 100, the underwriting operation is more than covering its costs.[7] Table 2.2 shows the evolution of the US Property and Casualty (P&C) insurance industry combined ratio since 2000.

With the exception of 2004 and 2006, this ratio has always been higher than 100 so that insurers did not collect sufficient premiums to cover losses

Table 2.2   US P&C insurance industry combined ratio, 2000–2006

	2000	2001	2002	2003	2004	2005	2006
Losses incurred	200.9	234.5	238.8	238.7	246.4	–	–
Loss adjustment expenses	37.8	40.9	44.8	50	53.2	–	–
Underwriting expenses	82.6	86.4	93.8	100.7	106.4	–	–
Earned premiums	294.0	311.5	348.5	386.3	412.6	–	–
Combined ratio	109	116	108	101	98	101	92.4

Source: Insurance Information Institute.

and expenses. For insurers to make a profit, their negative underwriting results have to be counterbalanced with sufficient returns from their investment portfolio management. After the severe 2005 hurricane season, there was only minor damage from disasters in 2006. The combined ratio for the year was 92.4, the lowest since 1949. Underwriting profit after dividends were $31 billion, only the third year that insurers had positive underwriting profits in the past 30 years in property/casualty coverage in the US.

HOW DOES THE US DEAL WITH DISASTER INSURANCE?

In the US, standard homeowners and commercial insurance policies, normally required as a condition for a mortgage, cover damage from fire, wind, hail, lightning, winter storms and volcanic eruption. Earthquake insurance can be purchased for an additional premium in all states except California where today one normally buys an earthquake policy for residential damage through the California Earthquake Authority, a state-run privately funded earthquake insurance program created in 1996. Earthquake coverage for businesses in California is often included in a commercial policy or can be purchased from private insurers as a separate rider. Flood insurance for residents and businesses is offered through the National Flood Insurance program, a public–private partnership created by Congress in 1968.[8]

Earthquake Insurance

Insurers provided coverage against earthquakes, floods and hurricanes without any public sector involvement until major disasters caused severe losses. In the case of earthquakes, the Northridge, CA earthquake of January 1994 caused $12.5 billion ($19.6 billion in 2007 prices; see Table

2.1) in private insured losses while stimulating considerable demand for coverage by residents in earthquake-prone areas of California. Insurers in the state stopped selling new homeowners policies that covered standard perils because they were also required to offer earthquake coverage as a rider to these policies for those who demanded it. This led to the formation of the California Earthquake Authority (CEA) in 1996 which raised the policyholders' deductible from 10 percent to 15 percent and limited the losses that insurers can suffer from a future earthquake (Roth, 1998). The CEA sells earthquake insurance policies for residential property throughout California. These policies are sold by insurers that participate in the CEA (22 insurers as of the end of 2005). These insurers act as independent contractor agents on behalf of the Authority by performing policy and claims services (underwriting, policy issuance, premium collection, claims adjustment and payment) and receive a portion of the collected claims to do so. As of 2007, the CEA had about 750 000 policies in place (the same number as in 2002 and 2005) (PricewaterhouseCoopers, 2006; www. earthquakeauthority.com/index.aspx?id=33).

Flood Insurance

Flood insurance was first offered by private companies in the late 1890s and then again in the mid-1920s. But the losses experienced by insurers following the 1927 Mississippi floods and severe flooding in the following year led all companies to discontinue coverage by the end of 1928 (Manes, 1938). Few private companies offered flood insurance in the next 40 years. Following Hurricane Betsy in 1965 which caused considerable damage to New Orleans, Congress passed the Southeast Hurricane Disaster Relief Act which provided up to $1800 in grants for those who suffered damage not covered by insurance. A study on the feasibility of flood insurance authorized by the Act reached the conclusion that some type of federal subsidy was required. The results of this study coupled with House Document No. 465 were instrumental in initiating the Congressional action which eventually led to the passage by Congress of the National Flood Insurance Program (NFIP) in 1968 (Kunreuther et al., 1978).

Today the federal government is the primary provider of flood insurance for homeowners and small businesses. Private insurers market coverage and service policies under their own names, retaining a percentage of premiums to cover administrative and marketing costs. In other words, they play the role of intermediaries between the insurers and the federal government.[9] Communities that are part of the program are required to adopt land use regulations and building codes to reduce future flood losses (Pasterick, 1998). In return, residents of these NFIP communities will be able to benefit from subsidized flood insurance premiums. Over time, the number of flood insurance policies has continuously increased; the program covered fewer than 2.5 million policies in 1990, 3.7 million in 2000, and 5.55 million as of 31 December 2007 (Michel-Kerjan and Kousky, 2008).

Wind Insurance

Coverage from wind damage is provided under standard homeowners and commercial insurance policies. Following Hurricane Andrew in 1992, which caused $23.7 billion in insured losses (in 2007 prices) to property on the southern coast of Florida, some insurers felt that they could not continue to provide coverage against wind damage in hurricane-prone areas within the state. This position was reinforced by insurance rate regulations that prevented them from charging the high rates that would be required to continue writing coverage. In response, the state formed the Florida Hurricane Catastrophe Fund (FHCF) in November 1993 (Lecomte and Gahagan, 1998). The FHCF acts as a state-administered reinsurance program providing reimbursements to insurers for a portion of their catastrophic hurricane losses and is mandatory for residential property insurers writing policies in the state of Florida. As of 2006, 205 insurers are covered by the fund. Although the FHCF is not backed by the state of Florida, it can call for emergency *ex post* assessments against all Florida property and casualty insurers (including surplus lines, but excluding medical malpractice until 1 June 2007, and workers' compensation) if its reserves are not sufficient to pay claims (State Board of Administration of Florida, 2005).

Major Limitations of these Programs

A closer examination of these insurance solutions reveals that these catastrophe programs were not set up to deal with a series of major catastrophes occurring over a short period of time such as the four hurricanes that devastated parts of Florida over six weeks in August and September 2004. Even though the FHCF paid less than 10 percent of the $22 billion in insured damage from these disasters, that was enough to put it under severe stress, and after the 2005 hurricane season it had a deficit of $1.6 billion. Most of the deficit will be funded through a $1.35 billion bond offering supported by a 1 percent assessment for six years on all property and casualty premiums (other than workers' compensation and medical malpractice). In 2007, the FHCF had an exposure of $2 trillion, 80 percent of which was residential coverage.[10]

On the earthquake side, the current mechanism in place does not really fit the potential for large-scale disaster. Plausible scenarios of major quakes in California imply losses that could be $100 billion or more. The CEA is not designed to deal with such large-scale disasters, as the maximum coverage provided by the different layers of the program and CEA's assets at year-end 2005 was $7.3 billion (including $1.5 billion provided by reinsurance) (PricewaterhouseCoopers, 2006). Moreover, quake insurance take-up in California has continuously been decreasing: it went from 30 percent when the CEA was established in 1996 down to 14 percent at the end of 2003 (Risk Management Solutions, 2004). But even with such a low take-up rate, the CEA will not have the financial strength to pay all claims should a large earthquake inflicting $100 billion or $200 billion in property damage occur in Los Angeles or San Francisco.

On the flood coverage side, Hurricane Katrina revealed that the take-up rate in the region impacted by the disaster was much lower than it should have been. In the Louisiana parishes affected by Katrina, the percentage of homeowners with flood insurance ranged from 57.7 percent in St Bernard's to 7.3 percent in Tangipahoa. Only 40 percent of the residents in Orleans Parish had flood insurance (Insurance Information Institute, 2005). These low percentages are particularly striking since the NFIP requires that homes located in Special Flood Hazard Areas purchase insurance as a condition for federally backed mortgages.

Following Hurricane Katrina, Congress approved additional emergency borrowing authority for the NFIP, which did not have enough reserves to pay all its claims. In March 2006 legislation was passed allowing the NFIP to borrow $20.8 billion, up from a limit of $1.5 billion. The NFIP borrows from the US Treasury in emergencies and repays the loan over time from premiums collected. It is unlikely that the NFIP will be able to repay the Treasury any time soon since it only collects about $2.8 billion a year in premiums.

TWO PRINCIPLES OF FUTURE DISASTER INSURANCE PROGRAMS

Recent events have demonstrated that current disaster insurance programs are not sustainable if the country suffers repeated large-scale disasters in a very short time frame. Insurers who are restricted from charging rates reflecting risk because of state regulations will refuse to continue to provide coverage in high-risk areas. If the price of insurance provides an inaccurate signal of the risk, then people and firms will continue to live and locate their activities in high-risk areas and will not have any incentive to invest in cost-effective mitigation measures. Indeed, the fast-growing concentration of values at risk in highly exposed areas, combined with a chronic underinvestment in protective activities, constitutes the perfect recipe for more devastating disasters in the future. According to data from AIR Worldwide, at the end of 2004, the total value of coastal insured exposure (both commercial and residential) in Florida was nearly $2 trillion ($1.4 trillion for commercial exposure and $900 billion for residential exposure),[11] which represented 80 percent of the statewide insured exposure. According to the Insurance Service Office (ISO), a direct hit on Miami by Hurricane Andrew that would have cost $60 billion in 1992 would have cost $120 billion in damage in 2004 because of the huge increase in new construction there and much higher property values.

Following the 2004 and 2005 hurricane seasons that considerably raised the level of awareness, a large number of residents have not invested in cost-effective loss-reduction measures with respect to their property or undertaken emergency preparedness measures. In a survey of 1100 adults living along the Atlantic and Gulf Coasts undertaken in May 2006, 83 percent had taken no steps to fortify their home, 68 percent had no hurricane survival kit and 60 percent had no family disaster plan (Goodnough, 2006).

Below we characterize two principles of insurance against catastrophic risks and illustrate how they could be applied to protect property against damage from future disasters: (1) establishing risk-based premiums and incentives for mitigation; and (2) dealing with equity and affordability issues. We also highlight some of the challenges in implementing these principles.

Principle 1: Risk-Based Premiums and Incentives for Mitigation

Insurance premiums should reflect the underlying risk associated with the events against which coverage is provided. To illustrate this principle, imagine you are an inhabitant of a city exposed to the risk of hurricanes. Assume also that the probability, as determined by a series of risk assessments of the area being hit by a major hurricane that would destroy the homes and apartments is 1/100. The expected loss to each property from hurricanes would be $2000. If the insurer sets rates that reflect the actuarial risk, then the annual premium would be $2000 $(1+\lambda)$, where λ is a loading factor reflecting the costs to the insurer of marketing a policy, assessing the risk, the cost of capital, settling the claim if any, and making normal profits. If $\lambda = 0.5$, then a risk-based premium for every piece of property in the city would be $3000.

Providing a clear signal of the danger

From an economic perspective, a principal advantage of risk-based premiums is that they provide a clear signal to individuals and businesses of the dangers they face. Rather than paying the very high premiums associated with living in hazard-prone areas, residents may choose to locate in safer locations where the price of insurance is lower. Current state regulations, and local economic and political interests, seriously limit insurers' ability to set premiums that reflect the risk. More generally, highly subsidized premiums due to rate regulation, without clear communication on the actual risk facing the homeowner, encourages development of hazard-prone areas in ways that are costly to both the individuals who locate there as well as the rest of society who are likely to incur the costs of bailing out victims following the next disaster.

Dollar payment versus probability
There is an additional reason why insurance premiums can make individuals aware of the relative risks associated with locating in different areas. Empirical studies have revealed that individuals rarely seek out probability estimates in making their decisions, and that low probabilities are inherently difficult to comprehend. When explicit probabilities are given to decision makers they often do not use the information (Magat et al., 1987; Camerer and Kunreuther, 1989).[12]

People are much more likely to pay attention to dollar expenditures when making location decisions. If they have comparative data on insurance premiums in different regions and know that these are risk-based rates, they will be able to determine the relative safety of different areas. In a controlled experimental study on whether individuals can distinguish between probabilities or insurance premiums for low probability events, Kunreuther et al. (2001) found that it was necessary to present comparative information on high- and low-risk situations for people to judge how safe an area would be with respect to its risk.

Providing incentives for mitigation
Risk-based premiums also allow an insurer to offer a rebate if the policyholder has invested in mitigation measures that effectively reduce the expected losses from perils to which he or she is exposed. Suppose you could reduce property damage caused by a hurricane by bracing your roof trusses and installing straps or clips at a cost of $1500. If the annual probability of a hurricane causing damage to your house is 1/100 and the reduction in loss due to strengthening the roof in this manner is $50 000, then the expected annual benefit from roof mitigation to your house is $500, and a risk-based insurance premium with $\lambda = 0.5$ should be reduced by $750 (that is, $500(1.5)).

Imagine now that a bank offers you a 20-year home improvement loan of $1500 at a 10 percent annual rate of interest to make your roof more hurricane resistant; your annual loan payment would be $145. You are much better off financially with this investment since your annual savings for the next 20 years would be $605 (for example, $50 a month). The bank also feels that it is better protected against a catastrophic loss to the property and the insurer knows that its potential loss from a major disaster is reduced. The general public will now be less likely to have large amounts of their tax dollars going for disaster relief. This represents a win–win–win–win situation for these concerned stakeholders.

Now suppose that, contrary to the previous example, an insurer would only be allowed to charge the Smiths $500 for an insurance policy due to state regulations. Then the insurer would have no economic incentive to provide a premium discount for undertaking a mitigation measure. In fact, no insurer would want to market coverage to this family or any homeowner in the hurricane-prone area with similar risks because in the long run the insurer would lose money on each of these policies.

Principle 2: Dealing with Equity and Affordability Issues

In proposing a disaster insurance program that stands a chance of being implemented in the real world, it is necessary to recognize the tension between setting premiums that are based on the real exposure to potential disaster and the financial ability of residents in hazard-prone areas to buy coverage. This raises a key question as to how best to deal with low-income people. One also needs to understand the role that residual markets play in certain states. We address both these issues in this subsection.

Special needs of low-income homeowners

In the case of Hurricane Katrina, many residents who lived in New Orleans did not purchase property (or flood) insurance. Most of them apparently felt that they could not afford the insurance and/or believed that the likelihood of damage to their property from a future hurricane was so low that they were not worried about its occurrence.

This equity issue is not a new one. It influenced the design of the National Flood Insurance Program (NFIP) in 1968. There was great concern that if flood insurance rates were simply risk-based, then many residents in hazard-prone areas would be charged extremely high premiums for flood coverage and would not want to purchase it. Hence the program was developed with two layers – a subsidized rate for residents currently residing in hazard-prone areas, and an actuarially based rate for those who built or substantially improved their structures after the federal government provided complete risk information in the area through flood insurance rate maps (Pasterick, 1998; Dixon et al., 2006). The subsidized rate was also designed to maintain the property values of structures in flood-prone areas.[13]

There are some lessons to be learned from the experience of the NFIP that should guide the development of a future disaster insurance program. First, even with highly subsidized insurance prices, many people did not have flood coverage. For example, a recent RAND Corporation study conducted for the NFIP revealed that only about 50 percent of single-family homes in special flood hazard areas are covered by flood insurance despite the subsidized rate for existing homes in flood-prone areas. More specifically, there are two types of special flood hazard areas to consider: first, those where there is a mandatory purchase requirement (for example, a mortgage) and those where there is not such a requirement. While there is a lot of uncertainty associated with where this requirement really applies, one estimates that about 50 to 60 percent of single-family homes located in special flood hazard areas are subject to the mandatory purchase requirement. Among these, the compliance rate varies a lot by region and across the country the rate is around 75 percent. But flood insurance market penetration is much lower for homes that are not subject to this requirement: only 20 percent of them purchase insurance despite the subsidized rates (Dixon et al., 2006).

Second, with subsidized rates there are no economic incentives for residents in hazard-prone areas to invest in mitigation measures because they will not be given premium discounts. Property owners that have repeatedly suffered damage from floods have rebuilt their property in the same location and continue to receive subsidized insurance rates. A recent US General Accountability Office (GAO) (2006) study revealed that structures receiving flood insurance payments of $1000 or more over a ten-year period constitute less then 1 percent of the properties covered under the NFIP but involve approximately 25–30 percent of all claims under the program. To address this problem the Flood Insurance Reform Act of 2004 provides states and local communities with an additional $40 million a year for mitigating severe repetitive-loss properties such as buyouts, elevation or moving the house (King, 2006).

Based on the experience of the NFIP, one should not provide subsidized premiums to those residing in hazard-prone areas. To be more explicit, the subsidy should not be provided through insurance prices. The premium should reflect the real exposure, the insured should pay it, but low-income residents in hazard-prone areas should receive some type of grant or tax rebate from the state or federal government that would defer some of the costs of the policy. To illustrate one possible arrangement, suppose that the Smith family in New Orleans had inherited their home from generation to generation and that the family's annual income is $50 000 so that they can not afford to pay $3000 for coverage against damage from hurricanes. Rather than having state regulators set a maximum premium of $1500 for insuring homes like theirs, the family could be given an insurance voucher that must be used to buy homeowners coverage.

The program could be similar in spirit to the food stamp program. The magnitude of the voucher would be based on the income and assets of the resident. Homeowners could also be provided with subsidies or loans to invest in cost-effective mitigation measures and in return be charged a lower insurance premium reflecting the reduced damage to their structure from a future disaster. The funds for supporting such a program could be raised through a surcharge on the property tax of all residents in the state or a surcharge on the property–casualty insurance policies sold throughout the state.

Residual markets and equity issues: the case of Florida[14]

Some states address the insurance affordability problem by giving all residents – not just those with low or moderate incomes – the opportunity to purchase property coverage, at subsidized rates, through the residual market, an insurer of last resort. Insurers operating in the residual markets are allowed to offer much lower prices, but under the condition that the homeowner has been rejected from coverage by at least one insurer. The residual market recoups the portion of insured losses that exceeded its reserves by levying an *ex post* surcharge against all insurance companies operating in the state (residential and commercial coverage). This surcharge is applied against all the policyholders. It raises a major equity issue because some policyholders who might have decided to live or work in very low-risk areas will end up paying for those in the state who have made a risky choice to buy or rent a house in a high-risk zone.

The residual market in Florida is operated by Citizens Property. Citizens was intended to be the insurer of last resort. But as a result of the market's reaction to the 2004 and 2005 hurricane seasons, which have inflicted $35 billion in total insured paid losses in Florida and obliged insurers to limit exposure or even stop covering high-risk areas, as of July 2006 Citizens had become the biggest insurer in this state. Because eligibility is not based on income or wealth but rather the inability of residents to purchase privately supplied insurance, Citizens has ended up covering high- and low-income policyholders alike. The 2004 and 2005 hurricane seasons have demonstrated that such insurance premium subsidies can produce major financial problems for the state. Because Citizens charged low premiums for properties in high-risk areas, its reserves were not sufficient to compensate all its policyholders. The four storms in 2004 cost Citizens $1.4 billion, the company was almost bankrupt at the end of 2004. To compensate for its $515 billion deficit, all personal and commercial property policyholders in Florida were assessed an additional 7 percent in 2005. Citizens' board also approved a filing of a 16 percent increase for those homes that it insured in high-risk areas.

As a result of the 2005 hurricane season, Citizens' deficit grew by another $1.77 billion. To limit this crisis, Florida's Governor Bush signed a property insurance reform bill that provided $715 million into Citizens in order to reduce this deficit. As a result, policyholders will face only an additional 2.5 percent surcharge rather than the maximum 10 percent permitted by law. The remaining deficit will be funded through a $1 billion debt offering to be paid off over ten years by additional policyholder assessments. Citizens is utilizing other sources of financing: it completed a $3 billion bond sale that increased its reserves (BestWire, 2006).

Moreover, the company benefits from a reinsurance arrangement with the Florida Hurricane Catastrophe Fund that would be difficult to obtain from the current market for catastrophe reinsurance. Indeed, according to the *Palm Beach Post*, Citizens has elected to forego available $500 million of private reinsurance should its losses exceed $5 billion because of the relatively high cost of $137 million for this coverage. Rather, Citizens purchased in Spring 2006 $5 billion of coverage in excess of $1.35 billion for only $250 million from the Florida Hurricane Catastrophe Fund (FHCF) (Horvath, 2006). As pointed out above, the FHCF has its own financial problems so it is not clear what motivated its decision to provide Citizens with reinsurance at such a low price compared with what the private reinsurance market currently offers.

Today it is widely recognized that Citizens' rates are still much too low relative to the risk. An important step toward the development of the risk-based premium we have been advocating was the approval by the Office of Insurance Regulation of Citizens' requested actuarial rates; statewide the rates will be increased by 23 percent (BestWire, 2006). While it is not clear whether this increase will eventually represent the actuarial price, it will certainly help Citizens and make homeowners realize more the risk they are exposed to. Furthermore, Citizens receives much of its reinsurance coverage from the FCHF, which charges premium rates that are estimated to be about one-quarter to one-third the cost of private market reinsurance (GAO, 2007). The company has recently decided that homes and condominium units valued at more than $1 million will not be covered by Citizens after July 2008, so these property owners will have to obtain coverage in the private insurance market or from surplus lines insurance companies whose rates are unregulated. While this new measure will certainly reduce Citizens' exposure, it will not be sufficient to stabilize the system (see Wharton Risk Center (2008) for a detailed discussion of the Florida market).

The example of Florida illustrates the current challenge faced by several states when it comes to defining and implementing a sustainable and equitable program to deal with natural disasters. As part of an ongoing initiative between the Wharton School, the Information Insurance Institute and Georgia State University, we are conducting a series of empirical analyses to determine what risk-based premiums would be in different states and specific metropolitan areas facing natural hazards as well as the pros and cons of alternative insurance programs that could be implemented at a state or national level.

WILL RISK-BASED ALL-HAZARDS DISASTER INSURANCE HELP IN THE POST-KATRINA WORLD?

This section considers the advantages and challenges of risk-based insurance that covers all hazards in a single policy. As discussed above, current insurance programs for residents in hazard-prone areas in the United States are segmented across perils. Standard homeowners and commercial insurance policies, normally required as a condition for a mortgage, cover damage from fire, wind, hail, lightning, winter storms and volcanic eruption. Earthquake insurance can be purchased for an additional premium. Flood coverage for residents and businesses is offered through the National Flood Insurance program, a public–private partnership created by Congress in 1968.

If one is to develop a comprehensive disaster insurance policy, it should adhere to the two principles of risk-based rates, and equity and affordability, discussed in the previous section. It enables insurers to diversify their risks and provides policyholders with the certainty that they are covered against all hazards. It does present challenges for small insurers that market policies in only a single state. All insurers have to convince a policyholder living far away from any water that he is not being charged a premium to cover the losses from those at risk from flood damage.

The idea of disaster insurance where all natural hazards are included in a single policy is not new. One of us proposed such a program for the United States many years ago (Kunreuther, 1968). In Europe, several countries have already adopted such a policy at a national level.

In 1954 Spain formed a public corporation, the Consorcio de Compensacion de Seguros, that today provides mandatory insurance for so-called 'extraordinary risks' that include natural disasters and political and social events such as terrorism, riots and civil commotion. Such coverage is an add-on to property insurance policies that are marketed by the private sector. The Consorcio pays claims only if the loss is not covered by private insurance, if low-income families did not buy insurance and/or the insurance company fails to pay because it becomes insolvent. The government collects the premiums and private insurers market the policies and handle claims settlements (Freeman and Scott, 2005).

In France, a mandatory homeowners policy also covers a number of different natural disasters along with terrorism. The main difference comes at the reinsurance level, which covers half of the insured losses suffered by the insurers and is provided by a publicly owned reinsurer, the Caisse Centrale de Reassurance (CCR), for flood, earthquakes and droughts. But insurers cannot benefit from this public reinsurance for wind damage. The French example is interesting because the country went through a similar debate several years ago on the wind–water controversy that is now occurring in the United States.

On 26 and 28 December 1999, Europe was devastated by two major windstorms. Within two days, these two storms killed 200 (the death toll would certainly have been much higher if the storms had occurred during a time when people were not staying with their families for Christmas). The storms also inflicted $10 billion in insured losses, making this event the most costly European catastrophe in the history of insurance. France was the country that suffered the highest damage. The debate occurred at an insurer–CCR level. The CCR had to reimburse insurers for the portion of their losses caused by flood; insured damage caused by wind was not reimbursed by the CCR. Despite this episode, the wind–water dichotomy is still an issue in France (Michel-Kerjan and de Marcellis, 2006).

Advantages of Risk-based All-hazards Disaster Insurance

Greater risk diversification
Consider an insurer marketing homeowners coverage in different parts of the country. With risk-based rates the insurer will collect premiums that reflect the earthquake risk in California, hurricane risk on the Gulf Coast, tornado damage in the Great Plains states and a flood risk in the Mississippi Valley. Each of these disaster risks is independent of the others. Using the law of large numbers, this higher premium base and the diversification of risk across many hazards reduces the likelihood that such an insurer will suffer a loss that exceeds the surplus in any given year for a given book of business.

Reduce uncertainty regarding causality
An all-hazards homeowners policy should also be attractive to both insurers and policyholders in hurricane-prone areas because it avoids the costly process of having an adjuster determine whether the damage was caused by wind or water. This problem of separating wind damage from water damage was a particularly challenging one following Hurricane Katrina. Across large portions of the coast, the only remains of buildings were foundations and steps where it proved is difficult to determine the cause of damage. In these cases insurers may decide to pay the coverage limits rather than

incurring litigation costs to determine whether the damage came from water or wind. For a house still standing, this process is somewhat easier since one knows, for example, that roof destruction is likely to be caused by the wind and water marks in the living room are signs of flooding (Towers Perrin, 2005).

An all-hazards policy would also deal with the problem that insurers currently face with respect to fire damage caused by earthquakes. Even if a homeowner has not purchased an earthquake insurance policy, he or she will be able to collect any damages from an earthquake due to fire. In the case of the 1906 San Francisco earthquake most of the damage was caused by fire and insurers were liable to cover these losses. In this sense, homeowners insurance actually covers a portion of earthquake losses even though this coverage is excluded from the policy.

Certainty of having coverage

Another reason for having an insurance policy that covers all hazards is that there will be no ambiguity by the homeowners as to whether or not they have coverage. Many residing in the Gulf Coast believed they were covered for water damage from hurricanes when purchasing their homeowners policies. In fact, lawsuits were filed in Mississippi and Louisiana following Katrina claiming that homeowners policies should provide protection against water damage even though there are explicit clauses in the contract that excludes these losses (Hood, 2005).

The attractiveness of insurance that guarantees the policyholder will have coverage against all losses from disasters independent of cause has also been demonstrated experimentally by Kahneman and Tversky (1979). They showed that 80 percent of their subjects preferred such coverage to what they termed probabilistic insurance where there was some chance that a loss was not covered. What matters to an individual is the knowledge that they will be covered if their property is damaged or destroyed, not the cause of the loss. Such a policy has added benefits to the extent that individuals are unaware that they are not covered against rising water or earthquake damage by their current homeowners policy.

A related advantage of a comprehensive homeowners program is that it may address some of the issues that currently plague the National Flood Insurance Program. As pointed out above, only half of the properties eligible for flood insurance are covered by it. Furthermore there were a number of properties suffering water damage from Hurricane Katrina that were not eligible to purchase flood insurance under the NFIP. Those who did have flood insurance and suffered large losses from the rising waters were only able to cover a portion of their losses because the maximum coverage limit for flood insurance under the NFIP is $250 000 on building property and $100 000 on personal property (Hartwig and Wilkinson, 2005a; 2005b).

Naturally, an all-hazards insurance policy might be more expensive than the policy a homeowner currently has. But if premiums are based on risk then policyholders would only be charged for hazards that they really face. Thus a homeowner in the Gulf Coast would theoretically be covered for earthquake damage but would not be charged anything for this additional protection if the area in which they reside is not a seismically active area. In promoting this all-risks coverage one needs to highlight this point to the general public who may otherwise feel that they are paying for risks that they do not face.

Challenges Associated with Risk-based All-hazards Disaster Insurance

Political issues associated with risk-based premiums

The major short-term disadvantage of a comprehensive disaster insurance program with risk-based rates is that it will force state regulators to raise their rates considerably to cover the potential damage in hazard-prone areas. For example, in Florida insurance rates along the coast subject to hurricanes are currently well below the actuarially fair premium (Grace et al., 2004). If insurance commissioners allow companies to charge a rate that reflects the risk, many individuals will be forced to pay premiums that are considerably higher than what they are currently charged. Many are likely to complain that this is highly unfair and unanticipated.

A large increase in premium will most likely be viewed by homeowners as unjustified and there will be significant resistance to paying for this coverage. For high-income residents who have second homes on the coast there is an economic rationale for them to pay the cost of their insurance. A step in this direction was recently discussed by the Florida legislature indicating that homes valued at over $2 million would have to be turned down by three surplus lines carriers (whose rates are not regulated) before they could turn to the state fund for coverage. For lower-income residents there needs to be an insurance subsidy from the state or federal government (as discussed above) so that these homeowners can afford to purchase coverage. Note, however, that this political problem arises if the insurance premium is risk-based for any peril.

Higher potential losses in a given period of time for insurers

Many insurers are likely to resist a comprehensive disaster insurance program because they may fear the possibility of even larger losses than they have suffered to date. Some note that if both wind and water damage were to be included in homeowners policies then the losses from a hurricane like Katrina to private insurers would be considerably higher. In order for insurers to feel comfortable with such a program they would have to be able to protect themselves against catastrophic losses either through private risk transfer instruments (for example reinsurance, catastrophe bonds), state funds or federal reinsurance.

Special treatment for small insurers

There will also be special needs facing small companies operating in a single state who have smaller surpluses than larger firms and are limited in their ability to diversify their risk. These insurers may find that the variance in their losses increases by incorporating the flood and earthquake risks as part of a homeowners policy. For example, a Louisiana insurance company providing protection against hurricane damage might find the variance in losses to be higher than it is today if both wind and water damage were covered under a homeowners policy. For these companies to compete with larger firms they would have to be able to protect themselves against catastrophic losses through either private or public-based risk transfer instruments that would not price them out of the market.

These smaller firms need to be differentiated from single-state subsidiaries established by some national insurer groups to help the parent company maintain or establish a high financial rating. The parent company has the option to disown itself from this single-state subsidiary should it suffer a catastrophic loss (Grace et al., 2006). There needs to be some protection given to the policyholders in this case should the single-state subsidiary declare itself to be insolvent.

Consumers' understanding of risk-based premiums

Insurers who market a comprehensive disaster insurance policy face an additional challenge in trying to convince homeowners that they are only paying for risks that they actually face. One way for them to do this is to itemize the cost of different types of coverage on the policy itself in much the way current homeowners or automobile insurance breaks up the cost for different types of protection. If the Smith family knew that it would be paying $3000 for wind coverage, $1500 for water coverage, $500 for fire coverage and $0 for earthquake coverage, it would not complain about covering damage from seismic risk facing California homeowners. Such an itemized list of coverage would also highlight the magnitude of risks that the Smiths faced by living in their home, another role that insurance can play – a signal as to how hazardous a particular place is likely to be.

Making disaster insurance mandatory

If insurance coverage against natural disaster is voluntary, it seems likely that the combination of reluctant insurers and reluctant buyers will result in limited coverage. Banks already require insurance coverage as a condition for loans and mortgages to protect their own financial interests. States or the federal government could require all residents and firms to purchase comprehensive disaster insurance.

Mandatory coverage would address many difficult problems. Since all residents would be financially protected, it would reduce the demand for government aid that is sure to arise after another disaster. The 2005 hurricane season has clearly illustrated that the federal government will always come to the rescue. The increase in the number of presidential disaster declarations since 1955 is illustrative of a growing involvement by the federal government in the recovery process. There were a total of 162 declarations between 1955 and 1965, 263 between 1976 and 1985, and 545 between 1996 and 2005 (Michel-Kerjan, 2008, forthcoming). It is not clear today how the large amount of federal disaster assistance post-Katrina will affect future behavior of individuals and firms located in hazard-prone areas.

If the prevailing view is that those residing in hazard-prone areas should be responsible for covering their own losses, then a mandatory insurance program would be appropriate. Many states require automobile insurance as a condition for obtaining a license. Disaster insurance could be treated in a similar fashion by including the premium as part of a person's property tax assessment. Such a proposal may be given serious consideration once one realizes that even after Katrina people are still not purchasing insurance in high-risk areas. One of the open questions, and one for which we need better data, is the number of uninsured homes in hazard prone areas, and why these people decided to be uninsured, so that we can determine ways to increase the take-up rate through government subsidy programs if this is a low-income issue.

CONCLUSION

The 2004 and 2005 hurricane seasons in the Gulf Coast have provided additional evidence that we have entered a new era in catastrophe management. Whilst the years 2006 and 2007 were relatively quiet ones on the natural disaster front, we can expect more frequent large-scale and devastating extreme events in the coming years. As part of the homeland security effort to protect citizens and the economy against untoward events (natural disasters, technological risks and terrorism threats), insurance will likely play a critical role not only in the recovery process afterwards, but also, if the price of coverage is related to risk exposure, in enhancing mitigation. If we as a society are to commit ourselves to reducing future losses from natural disasters and limiting government assistance after the event, then we have to modify current disaster insurance programs.

We proposed two principles for addressing this issue: (1) establishing risk-based premiums and incentives for mitigation; and (2) dealing with equity and affordability issues. Insurance with rates reflecting risk provides a signal of the hazardousness of the area and induces individuals to invest in adequate mitigation measures that will limit the consequences of these disasters. Long-term mitigation loans will provide the appropriate economic incentives for individuals to take these measures even if they have short-term planning horizons. There is an affordability issue that needs to be addressed through some type of subsidy given to low-income residents in hazard-prone areas so they are able to purchase insurance. But this subsidy should not be provided through subsidized insurance premiums.

Risk-based all-hazards disaster insurance is a natural candidate to consider when exploring ways to move from the status quo. While implementing such a program raises several challenges, we feel it is worth exploring in more detail through future research based on empirical data.

NOTES

1. For a more detailed analysis of the potential and challenges of a comprehensive disaster insurance program than the one provided in this chapter, see Kunreuther (2006).
2. Munich Re and Swiss Re, the two leading reinsurers in the world, do not use the same definition of catastrophic losses. Natural disasters inflicting insured losses above $38.7 million or total losses above $77.5 million are considered major catastrophes by Swiss Re. Munich Re considers a higher threshold. For example, when Munich Re estimated insured loss from natural disasters at about $42 billion in 2004, Swiss Re's estimate was over $52 billion. As a result, most figures used in the literature regarding the evolution of catastrophe loss actually underestimate the real effect on insurers.
3. This figure excludes payment by the US National Flood Insurance Program (NFIP) for damage from 2005 flooding (more than $20 billion in claims).
4. Poe financial group had three insurers that were liquidated in June 2006: Atlantic Preferred Insurance Co., Southern Family Insurance and Florida Preferred Insurance Co. Combined, the three companies had 320 000 policies, with a majority in South Florida. According to the Bermuda Monetary Authority, three months after Katrina hit the landfall, 11 new insurance companies had been created in Bermuda (for a total of $8.5 billion new capital).
5. The two most costly industrial catastrophes with respect to insurance claims over this 35-year period were the explosion of the oil offshore platform Piper Alpha in the North Sea in July 1988, which cost about $3.2 billion, and the explosion of a petrochemical plant in the US in October 1989 that inflicted insured damage of $2 billion (2004 indexed price).
6. See Kunreuther and Michel-Kerjan (2007) for a discussion on the question of attribution.
7. Even if the combined ratio exceeds 100 the insurer may still be making a profit based on the investment income from the premiums collected at the beginning of the insured period.
8. For more details on each of these insurance programs see Kunreuther and Roth (1998).
9. The NFIP is administered by the Federal Emergency Management Agency (FEMA) that is now part of the Department of Homeland Security's Emergency Preparedness and Response Directorate. In 2004, President Bush signed into law the Flood Insurance Reform Act, reauthorizing the NFIP through 2008.

10. We appreciate discussions with Jack Nicholson, President of the Florida Catastrophe Hurricane Fund, as well as his sharing exposure data with us.
11. As of 31 December 2005, current residential exposure for the state of Florida was just over $1.3 trillion (including $36 billion for mobile homes).
12. In one study, researchers found that only 22 percent of subjects sought out probability information when evaluating several risky managerial decisions. When another group of respondents was given precise probability information, less than 20 percent mentioned the probability in their verbal protocols (Huber et al., 1997). In other words, people do not deal with uncertainty in ways that would be predicted by normative models of choice.
13. The distribution of flood insurance business written in 2005 is anticipated to be 26 percent at subsidized rates and 74 percent at risk-based rates. Those being charged a subsidized rate are estimated to pay between 35 percent and 40 percent of the risk-based premium (Hayes and Sabada, 2004).
14. We appreciate insightful discussion with Jason Schupp from Zurich on the current status of the Florida residual market.

REFERENCES

Auerswald, Philip, Philip Branscomb, Todd LaPorte and Erwann Michel-Kerjan (eds) (2006), *Seeds of Disasters, Roots of Response. How Private Action Can Reduce Public Vulnerability*, New York: Cambridge University Press.
BestWire (2006), 'Florida's last resort insurer ranks first in policyholders', *AM Best*, 5 July.
Camerer, Colin and Howard Kunreuther (1989), 'Decision processes for low probability events: policy implications', *Journal of Policy Analysis and Management*, **8**, 565–92.
Department of Homeland Security (DHS) (2006), *National Infrastructure Protection Plan* (NIPP), 30 June, Washington, DC.
Dixon, Lloyd, Noreen Clancy, Seth Seabury and Adrian Overton (2006), 'The national flood insurance program's market penetration rate: estimates and policy implications', Report prepared by RAND for the American Institute for Research, Santa Monica, CA, February.
Freeman, Paul and Kathryn Scott (2005), 'Comparative analysis of large scale catastrophic compensation schemes', in OECD (2005), *Catastrophic Risks and Insurance*, Paris: Organization for Economic Cooperation and Development.
GAO (2006), *Federal Emergency Management Agency: Challenges for the National Flood Insurance Program*, Washington, DC: Government Accountability Office, 25 January.
GAO (2007), *Natural Diasasters: Public Policy Options for Changing the Federal Role in the Natural Castastrophe Insurance*, Washington, DC: Government Accountability Office, 26 November.
Goodnough, Abby (2006), 'As hurricane season looms, state aims to scare', *New York Times*, 31 May.
Grace, Martin, Robert Klein and Paul Kleindorfer (2004), 'Homeowners insurance with bundled catastrophe coverage', *Journal of Risk and Insurance*, **7**, 3 September.
Grace, Martin, Robert Klein and Zhiyong Liu (2006), 'Mother Nature on the rampage: implications for insurance markets', Paper presented at the NBER Insurance Workshop, Cambridge, MA, February.

Grossi, Patricia and Howard Kunreuther (eds) (2005), *Catastrophe Modeling: A New Approach of Managing Risk*, New York: Springer.

Hartwig, Robert and Claire Wilkinson (2005a), *Public/Private Mechanisms Handling Catastrophic Risk in the United States*, New York: Insurance Information Institute, October.

Hartwig, Robert and Claire Wilkinson (2005b), *The National Flood Insurance Program*, New York: Insurance Information Institute, October.

Hayes, Thomas and Shama Sabada (2004), *National Flood Insurance Program: Actuarial Rate Review*, Washington, DC: Federal Emergency Management Agency (FEMA), 30 November.

Hood, Jim (2005), 'A policy of deceit', *New York Times*, 19 November, p. A27.

Horvath, Stephanie (2006), 'Citizens pass up private reinsurance', *Palm Beach Post*, 26 May.

Huber, O., R. Wider and O. Huber (1997), 'Active information search and complete information presentation in naturalistic risky decision tasks', *Acta Psychologica*, **95**, 15–29.

Insurance Information Institute (2005), 'Flood insurance: facts and figures', 15 November.

International Panel on Climate Change (IPCC) (2001), 'WG2 Third Assessment Report, Chapter 8: Insurance and other financial services', New York: United Nations.

Kahneman, Daniel and Amos Tversky (1979), 'Prospect theory: an analysis of decision under risk', *Econometrica*, **47** (2), 263–91.

King, Rawle (2006), 'Hurricanes and disaster risk financing through insurance: challenges and policy options', Washington, DC: Congressional Research Service, 27 January.

Kunreuther, Howard (1968), 'The case for comprehensive disaster insurance', *Journal of Law and Economics*, April, **11**, 133–63.

Kunreuther, Howard (2006), 'Has the time come for comprehensive natural disaster insurance?', in R.J. Daniels, D.F. Kettl and H. Kunreuther (eds), *On Risk and Disaster*, Philadelphia: University of Pennsylvania Press.

Kunreuther, Howard and R.J. Roth, Sr (1998), *Paying the Price: The Status and Role of Insurance Against Natural Disasters in the United States*, Washington, DC: Joseph Henry Press.

Kunreuther, Howard and Erwann Michel-Kerjan (2007), 'Climate change, insurability of large-scale disasters and the emerging liability challenge', *University of Pennsylvania Law Review*, June, **55** (6), 1795–1842.

Kunreuther, Howard, Nathan Novemsky and Daniel Kahneman (2001), 'Making low probabilities useful', *Journal of Risk and Uncertainty*, **23** (2), 103–20.

Kunreuther, Howard, et al. (1978), *Disaster Insurance Protection: Public Policy Lessons*, New York: John Wiley & Sons.

Lecomte, Eugene and Karen Gahagan (1998), 'Hurricane insurance protection in Florida', in Howard Kunreuther and Richard Roth, Sr. (eds), *Paying the Price: The Status and Role of Insurance Against Natural Disasters in the United States*, Washington, DC: Joseph Henry Press, pp. 97–124.

Magat, Wes, W. Kip Viscusi and Joel Huber (1987), 'Risk-dollar tradeoffs, risk perceptions, and consumer behavior', in W. Viscusi and W. Magat (eds), *Learning about Risk*, Cambridge, MA: Harvard University Press, pp. 83–97.

Manes, Alfred (1938), *Insurance: Facts and Problems*, New York: Harper & Brothers.

Michel-Kerjan, Erwann (2006), 'Insurance, the 14th critical sector', in Auerswald, Branscomb, LaPorte and Michel-Kerjan (eds), *Seeds of Disaster, Roots of Response: How Private Action can Reduce Public Vulnerability*, New York: Cambridge University Press, pp. 279–91.

Michel-Kerjan, Erwann (2008), 'Toward a new risk architecture: the question of catastrophe risk calculus', *Social Research*, **75** (3) (forthcoming).

Michel-Kerjan, Erwann and Nathalie deMarcellis-Warin (2006), 'Public–Private programs for covering extreme events: the impacts of information distribution and risk sharing', *Asia-Pacific Journal of Risk and Insurance,* **1** (2), 21–49.

Michel-Kerjan, Erwann and Carolyn Kousky (2008), 'Come rain or shine: evidence on flood insurance purchases in Florida', Working paper, Wharton Risk Center, Wharton School; and Kennedy School of Government, Harvard University.

Munich Re (2006), 'Topics geo-annual review: natural catastrophes 2005', Knowledge series, report, Munich.

Pasterick, Edward (1998), 'The national flood insurance program', in Howard Kunreuther and Richard J. Roth, Sr. (eds), *Paying the Price: The Status and Role of Insurance Against Natural Disasters in the United States*, Washington, DC: Joseph Henry Press, pp. 125–54.

PricewaterhouseCoopers (2006), *California Earthquake Authority. Report on Audits of Financial Statements for the Years Ended December 31, 2005 and 2004*, report.

Roth, Richard Jr. (1998), 'Earthquake insurance in the United States', in Howard Kunreuther and Richard J. Roth, Sr. (eds.), *Paying the Price: The Status and Role of Insurance Against Natural Disasters in the United States*, Washington, DC: Joseph Henry Press, pp. 67–96.

State Board of Administration of Florida (2005), *Florida Hurricane Catastrophe Fund: Fiscal Year 2004–2005 Annual Report*, Florida.

Swiss Re (2005), 'Natural catastrophes and man-made disaster in 2004: more than 300 000 fatalities, record insured losses', *Sigma* no 1/2005, report.

Towers Perrin (2005), 'Hurricane Katrina: analysis of the impact on the insurance industry', http://www.towersperrin.com/tillinghast/publications/reports/Hurricane_Katrina/katrina.pdf

Wharton Risk Management and Decision Processes Center (2008), *Managing Large-Scale Risks in a New Era of Catastrophes: Insuring, Mitigating and Financing Recovery from Natural Disasters in the United States,* The Wharton School of the University of Pennsylvania, Philadelphia, March.

White House (2002), *National Strategy for Home land Security*, Washington, DC, July.

White House (2007), *National Strategy for Homeland Security*, Washington, DC: Homeland Security Council, Office of the President, October.

3. A decision analysis of options to rebuild the New Orleans flood control system

Carl Southwell and Detlof von Winterfeldt

INTRODUCTION

The levees and floodwalls protecting New Orleans from hurricanes and floods were designed to withstand a Saffir–Simpson category 3 hurricane (see US Army Corps of Engineers – USACE, 1984). When making landfall on 29 August 2005, Hurricane Katrina was designated a category 4 hurricane; later, it was downgraded to a severe category 3. The devastation that followed was more extensive than predicted by the USACE in 1984, but it was close to predictions made by scientists and emergency managers in more recent years (see Maestri, 2002; Laska, 2004). When examining the analyses conducted to support the 1984 decisions to fortify the levees and floodwalls, von Winterfeldt (2006, p. 31) concluded:

> In summary, there were several problems with the analyses and decisions regarding the development of levees and floodwalls in the New Orleans area: 1) probabilities and consequences of extreme hurricane events were underestimated; 2) alternatives that provided a higher level of protection were not explored; 3) the preferred alternative was implemented slowly and with many funding delays.

Subsequent reports (for examples, Interagency Performance Evaluation Task Force – IPET, 2006; Seed et al., 2006) came to similar conclusions.

More than a year later, the United States was again facing decisions about how to fortify and upgrade the flood protection system of New Orleans. In a previous paper (von Winterfeldt, 2006), we developed a simple decision tree analysis comparing two alternatives: rebuilding the levees and floodwalls to a 100-year flood protection level or building a new system that has a higher 1000-year protection level. Using a parametric analysis, the previous paper showed that a higher level of protection can be cost-effective. The previous paper also described improvements to be implemented in a more complete and comprehensive analysis.

This chapter makes another step in this direction by developing a decision analysis of options for the levee and floodwall system in and around New Orleans. Like the previous paper, we assume that substantial portions of New Orleans will be rebuilt and require protection. Moreover, we consider a comprehensive list of options for flood mitigation, of the possible types of events in terms of precipitation-, overtopping- and breach-induced floods, and of the consequences of these types of events. We use historical data to develop realistic estimates of flood frequencies and consequences and combine these estimates with a parametric analysis of events for which little historic data is available (for example, breaches, sabotage) or for which consequences are uncertain (for example, fatalities as a function of evacuation speed). We developed this analysis framework in the form of an influence diagram, a well-established modeling tool in risk and decision analysis (Clemen, 1997).

NEW ORLEANS' SYSTEM OF LEVEES AND FLOODWALLS

The levees and floodwalls developed by the USACE in the 1970s and 1980s reduced the risks of flood damage and provided economic development opportunities. At the time the USACE designed the system, its analysts believed that it protected New Orleans against a 100-year flood (that is, a flood of such magnitude that would occur, on average, only once in 100 years). However, due to many optimistic assumptions (for example, no levee breaches, rapid evacuation and resettlement, no consideration of fatalities), the analysts overestimated the level of protection and underestimated the consequences of such a major flood. In fact, the New Orleans area had experienced two near misses of category 3 hurricanes (Betsy in 1965 and Camille in 1969) which suggested that the probability of a category 3 or more severe hurricane (which would induce a '100-year' flood event) was much higher than one in 100 years.

Furthermore, the levels of protection decreased over time due to natural and man-made changes. Natural changes included continuing subsidence, lack of sedimentation and declining vegetative growth. Land use changes such as road building and increased residential densities induced hydrologic changes (including faster run-off) that reduced the level of protection provided by levees and floodwalls. And, while these levees and floodwalls required regular and extensive maintenance, their record of maintenance quality was spotty.

Over time, New Orleans' levees and floodwalls became structurally deficient and presented an increased risk to public safety and to the region's economic infrastructure. Minimum standards to regulate and to enforce the design, placement, construction and maintenance of levees and floodwalls had been and are critical to the built environment of New Orleans and its reconstruction. Indeed, the structural integrity and protection level of southeastern Louisiana's floodwall and levee system will strongly influence the extent of resettlement in New Orleans and influence the probability and consequences of future catastrophic hurricanes and floods.

In urban areas, the federal government has typically designed levees and other flood damage reduction projects with a 100-year flood threshold as the minimum standard for identifying, mapping and managing flood hazards. Participating National Flood Insurance Program (NFIP) communities are required to adopt building codes and other types of activities that reduce losses posed by a 100-year flood as a result of mandates by the Federal Emergency Management Agency (FEMA) and in order to maintain eligibility for the NFIP. FEMA also requires levees and floodwalls protecting flood-prone areas to be certified for structural soundness and proper maintenance to a 100-year flood level. The USACE performs most of these certifications. However, its current process does not assess the geotechnical or hydrological conditions of the levees, and neither the areas to be protected nor the structures built behind the protection of 100-year levees are classified as within 'designated floodplains'.

The accuracy of maps used by the FEMA to define flood hazard areas is also problematic, as more than three-quarters of these maps are more than a decade old, raising concerns that hydrologic data has changed since the maps were last reviewed and updated.

MODEL OVERVIEW

In modeling future floods and their expected consequences in New Orleans, many input quantities can only be estimated, and, as such, they have an inherent degree of uncertainty. A model that explicitly specifies the range of uncertainty in its inputs can provide more realistic and informative estimates than deterministic assessments. Influence diagrams are a useful tool in mapping out the decisions, events, and variables that influence the potential consequences of decisions and events (see, for example, Clemen, 1997). In this analysis, we use a software tool, Analytica (see www.lumina.com), to assist in modeling an influence diagram that represents the interrelationships among approximately 58 variables that include data for wind, rain, wave action, geology, engineering, demographics and the potential for negative consequences of hurricanes and floods in the New Orleans area.

At the highest level, we use a NOLA Flood Control Risk Analysis System with two major flood source submodels: Mississippi River flood frequency modeling (Model A) and Lake Pontchartrain flood frequency modeling (Model B), plus additional submodels that incorporate land use and mitigation options and demographic and consequence valuations for the New Orleans area. The submodels aggregate the expected frequencies of floods with their expected severities and present their expected costs as a function of their mitigation options. This model–submodel hierarchy of influence diagrams within Analytica serves as its key organizational tool. Because the visual layout of this influence diagram is intuitive, the model's stakeholders are able to learn about its structure and organization quickly through its visual paradigms.

The influence diagram also serves as a tool for communication. An understanding of how the results are obtained and of how the various assumptions impact the results is often more important than the specific input and output numbers. In addition to communicating high-level findings, stakeholders can examine lower levels of modeling when more detail is desired, aided by the visual aspects of the model's structure. As stakeholders are able to understand this model easily, debate and discussion can focus more directly upon specific assumptions and lead to more productive results. Thus, the influence diagram serves as a tool to help to make the model accessible.

Following is a brief description of the influence diagram structure, followed by a description of the model inputs and calculations.

The Mississippi flood submodel is shown in Figure 3.1. Floods are divided into two classes of chance nodes based on cause: overtopping and breaches caused by overtopping floods (which include upstream Mississippi River floodwaters compounded by sinking floodwalls and design errors as well as downstream Mississippi River surges compounded by sinking floodwalls and design errors); and breaches caused by anything other than overtopping (this includes terrorist acts, poor workmanship or materials, and design errors).

Figure 3.2 shows the Lake Pontchartrain flood submodel. Once again, floods are divided into two classes based on cause: overtopping and breaches caused by overtopping floods (which include Lake Pontchartrain surges, seiches and waves compounded by sinking floodwalls and design errors); and breaches caused by anything other than overtopping (this includes terrorist acts, poor workmanship or materials, and design errors).

The land use submodel includes the options considered in this analysis for improvements of the levee and floodwall system:

- Restoring the levees and floodwalls to their pre-Katrina conditions (base levels).
- Increasing the height of the levees and floodwalls by 5 feet.
- Increasing the height of the levees and floodwalls by 10 feet.

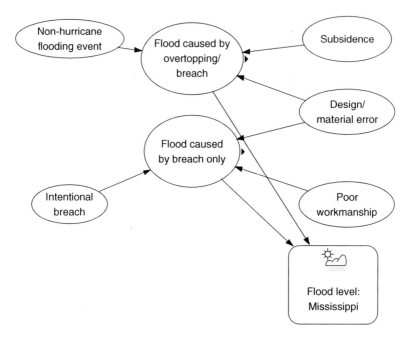

Figure 3.1   Mississippi flood submodel

Other options that can be explored with this model are improved levee maintenance and improved pumping systems and channels.

The demographics submodel contains the information representative of the housing stock and population in the New Orleans areas subject to possible flooding. This demographic information is used in the loss calculations, which determine, for each flood level, two consequences: lives lost and economic impacts. Lives lost are converted to economic equivalents by using a value of life of either $5 million or $10 million.

The analysis submodel (Figure 3.3) shows three decision nodes (rectangles). From the land use submodel, floodwall and levee heights can be selected. In addition, an option to allow the use of river flow cut-offs, such as the use of partial rechanneling of the Mississippi River down the Atchafalaya River during severe floods, is introduced. Attenuated by these choices, the products of flood and hurricane severities and frequencies return expected annual flood and hurricane losses and costs (net losses plus mitigation costs) for the New Orleans area. The uncertain quantities are specified using probability distributions. When evaluated, the distributions are sampled using Monte Carlo sampling, and the samples are propagated through the computations to the expected annual flood consequences (in

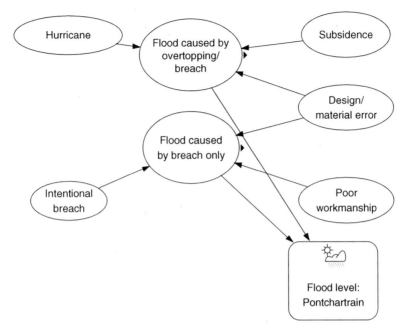

Figure 3.2   Lake Pontchartrain flood submodel

terms of lives lost and economic impacts). These distributions of consequences can then be analyzed in light of various mitigation strategies to
evaluate the expected costs and benefits of these strategies.

FLOOD FREQUENCY RISK ANALYSIS

The frequency distributions of potential floods in the New Orleans area
were based on historical data. As a starting point, we assumed that catastrophic floods could inundate New Orleans through two major pathways:
one primarily from the south and east via hurricanes as occurred with
Hurricane Katrina, and the other primarily from the north via Mississippi
River basin flood flows as in the extreme example of the Great Mississippi
Flood of 1927 (see Barry, 1997).

   In an attempt to capture accurate historic records of floods along the east
side of New Orleans (that is, the Lake Pontchartrain shoreline and similar
areas) and along the banks of the Mississippi River, we used United States
Geological Survey (USGS) flood gauge data of peak annual flood discharges
from a flood gauge station on Lake Pontchartrain and from two Mississippi
River gauges, one upstream in Baton Rouge, Louisiana, and one downstream

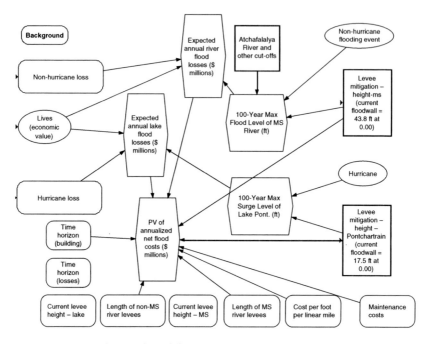

Figure 3.3   Analysis submodel

in West Pointe a La Hache, Louisiana. For the Mississippi River, two gauges were selected to represent maximum flood waters along the Mississippi River in New Orleans because, historically, cut-offs and intentional levee breaches have been used to temper rising waters along the banks of the Mississippi in New Orleans. The Baton Rouge station, then, was used as a proxy for maximum flood potential from Mississippi River basin floods that are resultant from upstream run-off, and the West Pointe a La Hache station was used as a proxy for downstream surge from approaching hurricanes.

Table 3.1 shows the relationship between storm categories, wind speed, minimum surface pressure and storm surge. Minimum pressures and surge heights are important in associating floods with the Saffir–Simpson scale of hurricane intensities (Simpson, 1974; also see http://www.ncdc.noaa.gov/oa/satellite/satelliteseye/educational/saffir.html).

For bodies of water with water-level gauges such as Lake Pontchartrain, a standard flood frequency analysis procedure is used. This procedure is promulgated from guidelines, known as *Bulletin 17* (U.S. Interagency Advisory Committee on Water Data, 1982), that are the official procedures of federal agencies in the United States. *Bulletin 17* characterizes flood

Table 3.1   Relationships between storm categories, wind speed, minimum surface pressure and storm surge

Saffir-Simpson Category	Wind speed		Minimum surface pressure	Storm surge
	mi/h	m/s	mb	ft
1	74–95	33–42	greater than 980	3–5
2	96–110	43–49	979–965	6–8
3	111–130	50–58	964–945	9–12
4	131–155	59–69	944–920	13–18
5	155+	69+	less than 920	18+

frequency at a given location as a function based on the sequence of annual data points known as the 'peak annual flood discharges' that are defined as the annual maximum water levels at the flood gauge location. These magnitudes are assumed to be independent random variables that are represented by log-Pearson Type III (gamma) probability distributions. These distributions give the annual exceedence probabilities, the probability that a flood will exceed a given magnitude in an annual period.

Bulletin 17 defines the annual peak flows for a site and describes the calculations in detail. Its steps include data collection, outlier detection and adjustment, skew adjustment, curve computation, plotting and confidence limits' calculation. A flood frequency curve is typically formulated for each type of hazard that is applicable, for example, upstream rainstorm, snowmelt run-off and hurricanes. As such, each hazard curve is a conditional probability curve. The unconditional probability distribution is obtained by weighting the conditional probability curves in proportion to the chance that a flood will be of each respective type. A means of expressing the magnitude of an expected flood is through the use of a term known as a 'return period' or probability of exceedence. The exceedence probability is not a random event, but a quantile of the flood frequency distribution. Thus, the probability of an exceedence next year for a 100-year return period is 1 percent, regardless of this year's outcome; the probability of exceedence in the year after next is 0.99×1 percent, and so forth, such that the average time to the next exceedence is 100 years.

The choice of a simple functional form for flood frequency distributions is problematic. Three of the more common choices for flood frequency are the extreme value distribution, the logistic distribution, and the lognormal distribution. We chose the logistic to represent the flood–surge exceedence curves for the Lake Pontchartrain floods and surges because it represents a reasonable fit to both the hurricane-induced and non-hurricane-induced floods, it is an available and flexible option within the Analytica modeling software, and its problematic tails are censored and truncated in the analysis. The logistic distribution's cumulative distribution function (cdf) is defined as follows:

$$F(x,\mu,s) = 1/(1 + e^{-(x-\mu)/s})$$

The probability density function (pdf) of the logistic distribution is given by:

$$f(x,\mu,s) = e^{-(x-\mu)/s}/[s(1 + e^{-(x-\mu)/s})^2]$$

The μ (mean) for the selected distribution is 10.487, and its s (shape) is 6.988. The fitted cdf was based on hurricane flood frequency calculations derived from standard project hurricane (SPH) frequency analyses for Lake Pontchartrain. This distribution represents the expected range of maxima of lake depths plus surge heights in feet over a return period equivalent to the number of years of data. An adjustment factor was used to convert the 32 available, annual data points to a distribution of measurements whose return period is 100 years (according to the formula, adjustment factor = (1 − [1/selected interval])/{1 − [1/actual interval]} or 1.021935484).

In addition to expected surge, seiche or wave maxima, the probability of non-overtopping-related breaches due to design errors, poor workmanship, improper materials and intentional sabotage as well as the gradual sinking of existing levees and floodwalls due to subsidence were incorporated. Design errors, poor workmanship and improper materials were estimated to cause catastrophic structural failure (without floodwater assistance) an average of once in 10 000 years. A Poisson distribution represents this failure rate. Intentional sabotage was estimated at a fixed probability of 1 in 10 000 per year, due to a lack of specific threat information. Average subsidence was estimated at 0.081 ft per year based on the estimates of subsidence as much as 0.162 ft per year. We also assumed that, once cumulative subsidence reaches 1 ft, mitigation occurs.

The product of these distributions returned a distribution of peak water levels for Lake Pontchartain (see Figure 3.4). The current average height of the levees and floodwalls above the lake's water level was estimated at 17.5 ft. From this measure, we constructed levee heights for different mitigation options at 17.5 ft, 22.5 ft and 27.5 ft. Floods are expected when peak water levels exceed the levee heights (represented as surge = 0 ft in the x-axes of Figures 3.4, 3.5a and 3.5b).

We determined the cumulative distribution function over flood levels in the Mississippi River in a similar manner. For purposes of this analysis, we selected the logistic function to represent the flood-surge exceedence curves

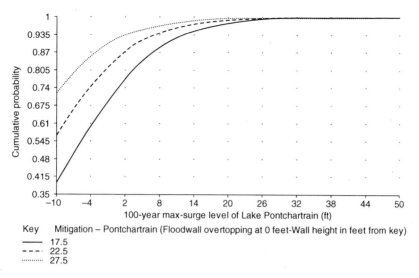

Key   Mitigation – Pontchartrain (Floodwall overtopping at 0 feet-Wall height in feet from key)
       ——— 17.5
       ---- 22.5
       ·········· 27.5

Figure 3.4   Cumulative distribution functions of 100-year flood levels for Lake Pontchartrain for different levels of protection

for the Mississippi River floods. The μ (mean) for this selected distribution is 23.483, and the *s* (shape) is 15.708. The fitted cdf was based on the distribution using non-hurricane flood frequency calculations starting with a standard log-Pearson Type III analysis for the Mississippi River at New Orleans, Louisiana representing river depth in feet over return period in years. This distribution represents the expected range of maxima of river depths over a return period equivalent to the number of years of data. An adjustment factor of 0.998182 was used to convert these 122 available annual data points to a distribution of measurements whose return period is 100 years.

In addition to expected floodwater maxima, the probability of non-overtopping-related breaches due to design errors, poor workmanship, improper materials and intentional sabotage as well as the gradual sinking of existing levees and floodwalls due to subsidence were incorporated. Design errors, poor workmanship, and improper materials were estimated to cause catastrophic structural failure (without floodwater assistance) an average of once in 10 000 years. A Poisson distribution represents this failure rate. Intentional sabotage was estimated at a fixed probability of 1 in 10 000 per year. Average subsidence was estimated at 0.081 ft per year based on the estimates of subsidence as much as 0.162 ft per year. We also assumed that, once cumulative subsidence reaches 1 ft, mitigation occurs.

The product of these distributions returned a distribution of peak water levels for the Mississippi River. The current average height of the levees and floodwalls above the river's bottom near its banks was estimated at 43.8 ft. From this measure, we constructed theoretical levee or floodwall heights at 43.8 ft, 48.8 ft and 53.8 ft (see Figure 3.5a). Floods are expected when peak water levels exceed the floodwall heights (represented as a flood when greater than 0.00 ft in the graph).

EVALUATION OF THE CONSEQUENCES OF FLOODS AND HURRICANES

We estimated both economic consequences of floods and the number of lives lost, depending on surge and flood levels, breaches and evacuation times. For the expected flood level for hurricanes, economic (excluding the value of lives) consequences were estimated in this analysis by utilizing historic economic consequences data collected by the National Oceanic and Atmospheric Administration (NOAA) (see Blake et al., 2006 and Landsea et al., 2003) adjusted to current levels (see Pielke et al., 2002). Historic hurricane losses were trended to current loss expectation levels by adjusting past losses for the cumulative effects of economic inflation, the growth of infrastructure and population change. The economic inflation adjustment was accomplished by using the annual Consumer Price Indices (CPI) from the US Bureau of Labor Statistics (see www.bls.gov). Infrastructure changes were quantified by using the annual indices measuring investments in fixed assets available from the US Bureau of Economic Analysis (see www.bea.gov). Finally, annual population estimates were derived from the US Bureau of the Census (see www.census.gov). The adjusted losses (from 1955 to current) were then fitted to a cumulative size-of-loss distribution as a gamma distribution (Figure 3.6) with an α of 0.1305 and a β of 62 500.

For non-hurricane floods, we estimated the non-hurricane flood economic (excluding the value of lives) consequences by utilizing historic economic consequences' data collected by the National Weather Service (NWS, a part of the NOAA) (see Pielke et al., 2002) adjusted to current levels (Figure 3.7). Historic flood losses were trended to current loss expectation levels by adjusting past losses for the cumulative effects of economic inflation, the growth of infrastructure and population change similar to the hurricane consequences' data. The adjusted economic losses (from 1955 to current) were then fitted to a cumulative loss distribution fitted as a log-logistic distribution (Figure 3.8) with a μ (mean) of 3.622 and an s (shape) of 2.996 (see Figure 3.8). In addition, we included an option for the use of cut-offs, such as the Atchafalaya River, during floods to decrease the peak flows. When the use of

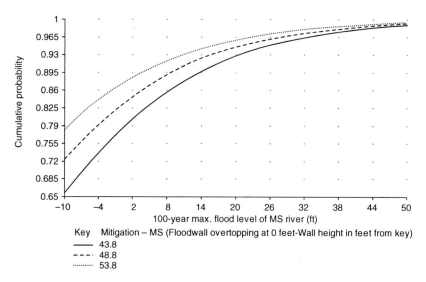

Figure 3.5a   Cumulative distribution function of 100-year flood levels at the Mississippi River for different protection levels and assuming no use of cut-offs

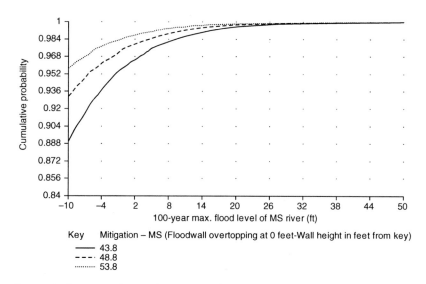

Figure 3.5b   Cumulative distribution function of 100-year flood levels at the Mississippi River for different protection levels and assuming use of cut-offs

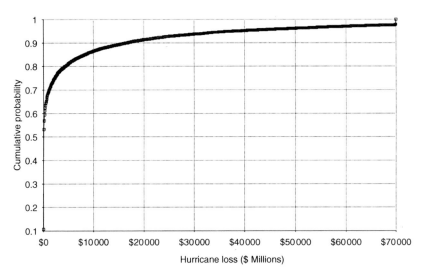

Figure 3.6   Cumulative severity distribution of hurricane losses (in $ millions and excluding value of lives)

cut-offs is allowed, it was assumed to reduce the floodwater peaks by 50 percent and vastly reduce the potential for a Mississippi River inundation in the city of New Orleans (see, for example, Figure 3.5b).

The economic value of losses of lives for both hurricane and non-hurricane floods was estimated in this analysis as a function of the population at risk, the evacuation time, and the assigned economic value of a lost life, namely, the economic value of lives lost equals the selected value of life times the estimated lives lost where the estimated lives lost is assumed to be the ratio of the population at risk of dying to the product of the estimated number of hours of evacuation time (before the flooding event) and 36.236466 (Stedge et al., 2006). Figure 3.8 uses $10 million as the value of a life and displays the economic values of lives lost as a function of evacuation time. Note that this is independent of the type of flooding event.

The aggregate economic value of an event, then, is derived simply as the products of the frequencies of these events and the sums of their independent economic severity distributions and the value of life distributions.

SOME PRELIMINARY RESULTS

We consider first a base case analysis, comparing the expected costs of several options to reduce the risk of floods in the New Orleans area,

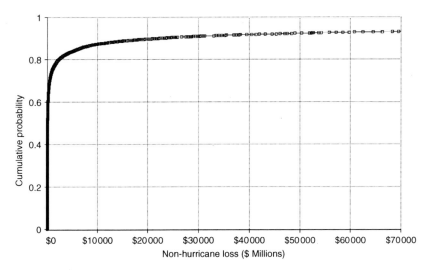

*Figure 3.7   Cumulative severity distribution of non-hurricane flood losses
(in $ millions and excluding value of lives)*

followed by several sensitivity analyses. Figure 3.9 shows, in the form of bar charts, how the expected costs compare to one another. There are three major messages conveyed by this figure. First, the expected consequences of a flood are dominated by the economic impacts rather than by the potential fatalities; second, the mitigation costs are commensurate with the economic costs of floods; and, third, there appear to be three contenders that minimize the total expected costs: the status quo with cut-offs, increased levee heights at the Mississippi with cut-offs, and increased levee heights at Lake Pontchartrain with cut-offs. Note that all three options include cut-offs.

Lake Pontchartrain mitigation options are substantially more expensive than Mississippi River options, but the savings in terms of economic losses avoided tend to more than make up for the expense. In this analysis, mitigation of floodwalls and levees by increasing height is assumed to cost $3 265 000 per vertical foot per mile. The Mississippi side of the levee and floodwall system is approximately 100 miles long; the Lake Pontchartrain side (including the interior fortifications), about 250 miles.

Interestingly, the Atchafalaya River and other potential 'relief valves' that serve as flood flow cut-offs in the event of upstream flooding provide an estimated mitigation value of as much as $2.3 billion annually. Indeed, New Orleans has depended on such cut-offs historically to avoid Mississippi River inundations. In the future, given the physical characteristics of the upstream region that is home to the Old River Control Structure and similar upstream

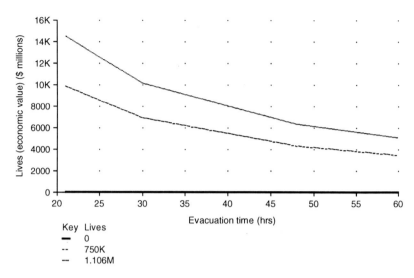

Figure 3.8   Cumulative severity of lives (in $ millions) as a function of evacuation time

areas with significant sandy deposits and gradients steeper than the current Mississippi River bed, it may be prudent to consider carefully this traditional mitigation strategy. Potentially, its use could catalyze the necessary initial conditions for the avulsion of the Mississippi to the Atchafalaya. An avulsion of the Mississippi River has the potential to doom irreparably the economy and future welfare of New Orleans and Baton Rouge.

This analysis is most sensitive, respectively, to the economic value we impute to a human life, to mandatory evacuation time, and to the combined levee or floodwall height on the Lake Pontchartrain side of New Orleans. Simply stated, the most immediate, significant flood and hurricane mitigations in New Orleans can be accomplished by increasing minimum, mandatory evacuation times for hurricanes to 48 hours or more and by giving first priority to repairs and fortifications of levees and floodwalls on the Lake Pontchartrain side of New Orleans.

CONCLUSIONS

To avoid repeating the mistakes of the past, this analysis can serve as an example of being more realistic about the assessment of probabilities of these future extreme events and about their consequences. Some preliminary results include the following:

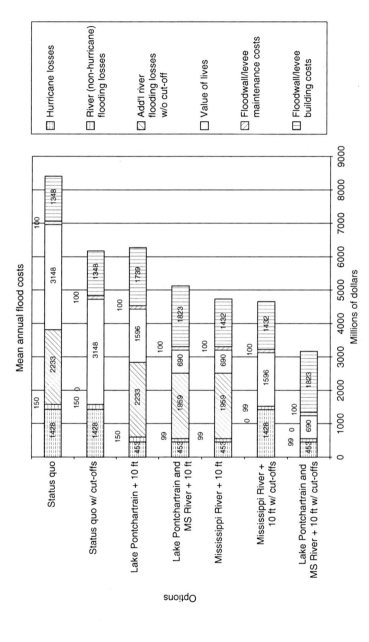

Mean annual flood costs

Legend:
- ▦ Hurricane losses
- ▤ River (non-hurricane) flooding losses
- ▨ Add'l river flooding losses w/o cut-off
- ☐ Value of lives
- ▧ Floodwall/levee maintenance costs
- ▦ Floodwall/levee building costs

Options (vertical axis):
- Status quo
- Status quo w/ cut-offs
- Lake Pontchartrain + 10 ft
- Lake Pontchartrain and MS River + 10 ft
- Mississippi River + 10 ft
- Mississippi River + 10 ft w/ cut-offs
- Lake Pontchartrain and MS River + 10 ft w/ cut-offs

Millions of dollars (horizontal axis): 0, 1000, 2000, 3000, 4000, 5000, 6000, 7000, 8000, 9000

Status quo: 428, 150, 2233, 150, 3148, 100, 1348

Status quo w/ cut-offs: 1428, 150, 0, 3148, 100, 1348, 100

Lake Pontchartrain + 10 ft: 455, 150, 2233, 100, 1596, 100, 1789

Lake Pontchartrain and MS River + 10 ft: 455, 99, 1959, 690, 100, 1823

Mississippi River + 10 ft: 455, 99, 1959, 690, 100, 1432

Mississippi River + 10 ft w/ cut-offs: 1428, 0, 99, 1596, 100, 1432

Lake Pontchartrain and MS River + 10 ft w/ cut-offs: 455, 99, 0, 690, 100, 1823

Note: These are from all causes assuming a population of 750 000, 48-hour evacuation time, mitigation costs for levee and floodwall fortification and maintenance, 7 percent interest, $3.265 million per vertical foot per mile construction costs, and an imputed value of $10 million per life.

Figure 3.9 Net annual flood costs

49

- Increasing the floodwall and levee heights by 10 feet can be cost-effective.
- Continuing to provide a Mississippi River cut-off option seems to be very cost-effective.
- Increasing mandatory hurricane evacuation periods to at least 48 hours and, on a prescribed basis, for up to 60 hours can be very cost-effective and save many lives.

This analysis can be improved in several important ways:

- Improving the breach and overtopping model.
- Using formal expert judgment methods to improve the assessment of probability distributions and their parameters.
- Using uncertainty analysis to account for changes in frequencies and severities of flooding due to climate change phenomena.
- Using uncertainty analysis to assess the impacts of subsidence and the benefits of subsidence mitigation options.
- Using uncertainty analysis to assess the feasibility and effects of varying evacuation times.
- Developing and analyzing a more complete set of consequence measures, including impacts on ecology, habitat, environmental justice, and so on.
- Involvement of stakeholders in the design and modification of the analysis and in the interpretation and communication of the results.

In addition, it may be worthwhile to consider other options, for example:

- Assigning floodplains that are enclosed by levees and floodwalls within prescribed bounds the status of 'designated floodplain', regardless of the engineering standards of the levees and floodwalls.
- Reworking floodplain maps on a regular basis to reflect more accurately elevation changes due to natural (for example, subsidence, erosion) and man-made impacts (for example, global warming, subsidence caused by oil and gas extraction, hardscaping effects).
- Addressing floodwall and levee improvement in creative ways (for example, considering use of slurry walls in levees, trading-off additional floodwall height for marsh restoration – for example 1 ft of floodwalls is approximately equivalent to 2.7 miles of restored marshlands – or other sustainable improvements).
- Considering the probable avulsion of the Mississippi River in considering the refortification and rebuilding of New Orleans.

Perhaps an optimal allocation would incorporate a multiple-lines-of-defense strategy that incorporates the principles and lessons of integrated coastal zone management (ICZM) and includes the combined buffering impacts of the offshore shelf within the Gulf of Mexico, the Louisiana barrier islands, the Louisiana sounds, marshland bridges, natural ridges, man-made soil foundations, floodgates, flood protection levees, flood protection pumping, elevated homes and businesses, and enhanced and more timely evacuation procedures (IPET, 2006).

Hurricane Katrina was a major natural disaster, the impacts of which were exacerbated by a poorly performing flood protection system due to engineering and institutional failures, questionable judgments, and errors involved in the design, construction, operation and maintenance of the system. The organizational and institutional problems associated with the response and recovery efforts for this combined natural and man-made disaster resulted in one of America's most severe catastrophes.

REFERENCES

Barry, John M. (1997), *Rising Tide: The Great Mississippi Flood of 1927 and How It Changed America*, New York: Touchstone Books.
Blake, Eric S., Jerry D. Jarrell, Edward N. Rappaport and Christopher W. Landsea (2006), *The Deadliest, Costliest, and Most Intense United States Tropical Cyclones from 1851 to 2005 (and Other Frequently Requested Hurricane Facts)*, *NOAA Technical Memorandum NWS TPC-4*, Miami, FL: NOAA, Updated July 2006, http://www.nhc.noaa.gov/Deadliest_Costliest.shtml.
Clemen, Robert T. (1997), *Making Hard Decisions: An Introduction to Decision Analysis*, Pacific Grove, CA: Duxbury Press.
Interagency Performance Evaluation Task Force (IPET) (2006), 'Performance evaluation of the New Orleans and Southeast Louisiana hurricane protection system: draft final report', United States Army Corps of Engineers (USACE), 1 June, see also https://ipet.wes.army.mil/.
Landsea, Christopher W., Craig Anderson, Noel Charles, Gilbert Clark, Jason Dunion, Jose Fernandez-Partagas, Paul Hungerford, Charlie Neumann and Mark Zimmer (2003), *The Atlantic Hurricane Database*, Re-analysis Project Documentation for 1851–1910 Alterations and Addition to the HURDAT Database, http://www.aoml.noaa.gov/hrd/hurdat/Documentation.html.
Laska, S. (2004), 'What if Hurricane Ivan had not missed New Orleans?' *Natural Hazards Observer*, **29**.
Maestri, W. (2002), Interview with American RadioWorks, http://american radioworks.publicradio.org/features/wetlands/hurricane2.html.
Pielke, Jr., R.A., M.W. Downton and J.Z. Barnard Miller (2002), *Flood Damage in the United States, 1926–2000: A Reanalysis of National Weather Service Estimates*, Boulder, CO: UCAR http://www.flooddamagedata.org/full_report.html.
Seed, R.B., R.I. Abdelmalak, A.G. Athanasopoulos, R.G. Bea, G.P. Boutwell, J.D. Bray, J.-L. Briaud, C. Cheung, D. Cobos-Roa, J. Cohen-Waeber, B.D. Collins, L. Ehrensing, D. Farber, M. Hanenmann, L.F. Harder, M.S. Inamine, K.S. Inkabi, A.M. Kammerer, D. Karadeniz, R.E. Kayen, R.E.S. Moss, J. Nicks, S. Nimala,

J.M. Pestana, J. Porter, K. Rhee, M.F. Riemer, K. Roberts, J.D. Rogers, R. Storesund, A. Thompson, A.V. Govindasamy, X. Vera-Grunauer, J. Wartman, C.M. Watkins, E. Wenk and S. Yim (2006), 'Investigation of the performance of the New Orleans flood protection systems in Hurricane Katrina on August 29, 2005', Report No. UCB/CCRM – 06/01, University of California, 22 May.

Simpson, R.H. (1974), 'The hurricane disaster potential scale', *Weatherwise*, 27,169,186.

Stedge, Gerald, Mark Landry and Maged Aboelata (2006), 'Estimating loss of life from hurricane-related flooding in the Greater New Orleans area – Loss-of-life modeling report: final report', Alexandria, Virginia: US Army Corps of Engineers Institute for Water Resources, 22 May.

Townsend, Frances Fragos, Kenneth P. Rapuano, Joel B. Bagnal, S. Michele, L. Malvesti, Kirstjen M. Nielsen, Thomas P. Bossert, Daniel J. Kaniewski, Marie O'Neill Sciarrone, Joshua C. Dozor, Michael J. Taylor, Stuart G. Baker, Richard W. Brancato, Donovan E. Bryan, Christopher Combs, Theodore M. Cooperstein, William T. Dolan, Michael O. Forgy, Douglas J. Morrison, Richard L. Mourey and David C. Rutstein (2006), *The Federal Response to Hurricane Katrina: Lessons Learned*, Washington, DC: White House, 23 February.

United States Army Corps of Engineers (USACE) (1984), *Lake Pontchartrain, Louisiana, and Vicinity Hurricane Protection Project: Reevaluation Study*, New Orleans, LA: US Army Corps of Engineers.

U.S. Interagency Advisory Committee on Water Data (1982), *Guidelines for determining flood flow frequency*, Bulletin 17-B of the Hydrology Subcommittee: Reston, Virginia, U.S. Geological Survey, Office of Water Data Coordination, Available from National Technical Information Service, Springfield VA 22161 as report no. PB 86 157 278 or from http://www.fema.gov/mit/tsd/dl_flow.htm.

von Winterfeldt, Detlof (2006), 'A risk and decision analysis framework to protect New Orleans against another Katrina', in R.J. Daniels, D.F. Kettl and H. Kunreuther (eds), *On Risk and Disaster*, Pittsburgh, PA: Pennsylvania State University Press.

4. Hurricane Katrina: lessons learned

Jiin-Jen Lee and Bennington Willardson

INTRODUCTION

New Orleans is an important port city in southern Louisiana. The city is a center of shipping for Midwest agricultural products, has a rich heritage and culture, and is a popular tourist destination. New Orleans is also below sea level and continues to sink lower as time passes.

Cradled in a wide southern meander of the Mississippi River just north of the Gulf of Mexico, New Orleans is surrounded by Lake Pontchartrain to the north, Lake Borgne to the east, and Lakes Cataouatche and Salvador to the south. This ring of fresh water is also surrounded by hundreds of square miles of wetlands and the Gulf of Mexico. The city has been protected from the river, lakes and ocean by construction of a levee system (Brouwer, 2003).

Hurricane Katrina, a category 4 storm, blew through the Gulf of Mexico and made landfall on 29 August 2005. Levees failed during the hurricane and the city of New Orleans was flooded. Hurricane Katrina was the largest natural disaster ever to occur in the history of the United States. The death count was estimated to be in the order of 1500 people. Many residents were stranded for days and weeks awaiting rescue, homes and businesses were destroyed, and damage to oil production facilities in the Gulf caused price increases throughout the nation. Estimates of the damages to the greater New Orleans area ran as high as $150 billion.

This chapter discusses the pre-storm conditions of the city and its protective levees, the effects of the storm, and the levee failures. The chapter focuses on the collective engineering lessons learned from Hurricane Katrina. We hope that insights from this devastating event will benefit city reconstruction efforts and be applied to the planning and design of systems that protect lives and property from natural and man-made disasters.

CONDITIONS PRIOR TO HURRICANE KATRINA

The French explorers Ibeville and Bienville founded New Orleans in 1718. The area was recognized as a strategic location for a city since it was at the outlet of the extensive network of rivers that form the Mississippi River drainage basin. The river network drains much of the land between the Rockies and the Appalachian mountain ranges.

New Orleans was ideal for portage because traders could bypass the treacherous lower 100 miles of river by crossing Lake Pontchartrain. However, there was no ideal location for a city at the mouth of the Mississippi River due to seasonal flooding and the low swampy land (Brouwer, 2003).

In the late 1800s, the United States Army Corps of Engineers (Corps) began constructing permanent levees along the river channel. With the levees in place, the lowlands beyond the river did not flood as often and people began building homes.

History of Flooding and Disaster Response

The Mississippi River flooded in 1927, killing at least 1000 people and inundating 1 million homes. Congress directed the US Army Corps of Engineers to straighten the river in places, add floodgates and increase the height of the levees from Vicksburg, Mississippi to the Gulf of Mexico. Today almost 200 miles of canals lead to 22 pumping stations located in the low points of the city. The stations are able to pump 35 billion gallons of water per day from the city into the surrounding lakes. The city is now well protected from the Mississippi River.

The New Orleans region has been extensively flooded by hurricanes six times in the last century: 1915, 1940, 1947, 1965, 1969 and 2005. On 9 September 1965, Hurricane Betsy hit the Gulf Coast on the southern tip of Louisiana. Almost every building in the small coastal town of Grand Isle was quickly destroyed. Betsy moved towards New Orleans with 150 mph winds. Lake Pontchartrain swelled with a storm surge that flowed through canals from the Gulf of Mexico. Easterly winds blew across the lake, causing large waves that resulted in water overtopping the levees.

After Hurricane Betsy, Congress appropriated funding to create a more advanced levee system to protect the city. Engineers determined that the Mississippi River levees were sufficient to withstand the surge produced by the Standard Project Hurricane (SPH) that was developed for design after Hurricane Betsy. The design concept for protecting the city was to create a rough semicircle of hurricane levees along the boundary of Lake Pontchartrain. The levee was to start at the Mississippi River levee west of New Orleans and end at a point east of the city. Levees and floodwalls would have to be constructed around drainage canals, bridges crossing the canals and the pump stations emptying into the lake (Brouwer, 2003).

The Corps began the large-scale project. As construction progressed, one project led to another. In 1986, Congress authorized the Corps to build a system of levees around the southern half of the city, in the form of a semi-circle between Mississippi River levees. Congress continued to expand the southern protection zone in the 1990s, requiring a total of about 65 miles of levees to protect thousands of homes. The scheduled date for completion of the entire levee system was 2018, 53 years after Hurricane Betsy emphasized the need for the improved hurricane protection system (HPS) (Brouwer, 2003).

Level of Protection Questioned Prior to Hurricane Katrina

Protection of the city by the levees was questioned in 2003. The design of the levee system was developed before advanced computer methods for storm prediction and modeling were developed. The system was also developed before the introduction of the Saffir–Simpson scale that is currently used by the National Weather Service to categorize hurricane intensity. The Saffir–Simpson scale is provided in Table 4.1 and ranks storms based on wind speeds, storm surge and expected damage (NWS, 2006).

The Corps predicted the storm's effects on the system based on the Standard Project Hurricane (SPH). They also computed the build-up of water at the levees along the lake's south shore. They ran simulations mimicking the characteristics of hurricanes that hit New Orleans in 1915 and 1947. Then they used the ratio between the actual storm surge recorded during those hurricanes and their computed value to create a factor of safety for the new levees.

Corps engineers determined that the SPH was equivalent to a fast-moving category 3 storm on the Saffir–Simpson scale. If a category 3 hurricane stalled over the city, or if a larger scale storm made landfall, analysis predicted that the levees would fail, leaving the city under 20 ft of water (Brouwer, 2003).

Some experts predicted that a less severe storm could flood the city since the design criteria were established for New Orleans' hurricane protection levees 40 years ago. The design did not account for subsidence of southeastern Louisiana's coastline causing the ocean level to rise relative to the city's ground level. The wetlands of southern Louisiana, the largest in the United States, which provide a buffer against storm surges and waves, are being lost at an unprecedented rate: approximately 1900 square miles since 1930 (Dean, 2006). Over the last century, the 50-mile wide marsh area between the city and the ocean has shrunk to a 25-mile wide marsh (Bourne, 2004).

Since the 1980s, computer models of the area have been created to predict the effects of the higher-category hurricanes on New Orleans. At least three

Table 4.1   Saffir–Simpson scale

Category	Maximum sustained wind speed (mph)	Minimum surface pressure (mb)	Storm surge (ft)	Damage description
1	74–96	> 980	3–5	Damage to unanchored mobile homes, shrubbery and trees. Some damage to signs.
2	97–111	979–965	6–8	Some roofing material, door, and window damage of buildings. Considerable damage to shrubbery and trees with some trees blown down. Considerable damage to mobile homes, poorly constructed signs, and piers.
3	112–131	964–945	9–12	Some structural damage to small residences and utility buildings with a minor amount of curtainwall failures. Damage to shrubbery and trees with foliage blown off trees and large trees blown down. Mobile homes and poorly constructed signs are destroyed.
4	132–155	944–920	13–18	More extensive curtainwall failures with some complete roof structure failures on small residences. Shrubs, trees, and all signs are blown down. Complete destruction of mobile homes. Extensive damage to doors and windows.
5	156+	< 920	19+	Complete roof failure on many buildings. Some complete building failures, including small utility buildings and mobile homes. All shrubs, trees and signs blown down. Severe and extensive window and door damage.

different models have been used to determine the effects in this area. The Advanced Circulation Model for Coastal Ocean Hydrodynamics (ADCIRC) is a finite element numerical model capable of handling 1 million nodes. It predicts long-term periods of circulation and water surface elevations along coastal shelves, coasts and estuaries in two dimensions (Brouwer, 2003).

In the 1980s, Joseph Suhayda at Louisiana State University (LSU) simulated storms with a modified version of a hurricane model used by the Federal Emergency Management Agency (FEMA) (Suhayda, 1985; Aravamuthan and Suhayda, 1999). Suhayda's model contains a geographic information system overlay that divides a fairly large boundary, from Alabama to Texas, into 0.6 mile (1 km) grids containing information about ground elevations, land masses and waterways. The FEMA hurricane model does not draw on the same processing power as the ADCIRC and produces more liberal projections of flooding from storm surges (Brouwer, 2003).

Suhayda was able to use the model to predict the storm surge associated with an actual hurricane dozens of hours before it hit land. By subtracting the elevations on a topographical map of coastal Louisiana from those surge values, he was able to approximate the flood risk of a given storm. Suhayda's data indicate that the water level of Lake Pontchartrain would rise by as much as 12 ft. As the storm's counterclockwise winds battered the levees on the northern shore of the city, the water would easily top the embankments and fill the streets to a depth of 25 ft or more (Brouwer, 2003).

The Sea, Lake and Overland Surges from Hurricanes (SLOSH) model is used by the National Weather Service and local agencies for emergency preparedness. The SLOSH model does not contain as many computational nodes as the ADCIRC model, does not use a finite-element grid to increase the resolution of the nodes on shore, and has a much smaller boundary. Even so, the SLOSH model indicates that the results would be devastating if a category 4 or 5 storm hit New Orleans (Brouwer, 2003).

EFFECTS OF HURRICANE KATRINA ON NEW ORLEANS

Hurricane Katrina made landfall between Gulf Port, Mississippi and New Orleans, Louisiana on 29 August 2005. The category 4 hurricane had sustained winds of 145 miles per hour and produced large onshore storm surges from the Gulf of Mexico. Figure 4.1 shows a map of the calculated storm surge levels at the time when the eye of the storm passed close to the

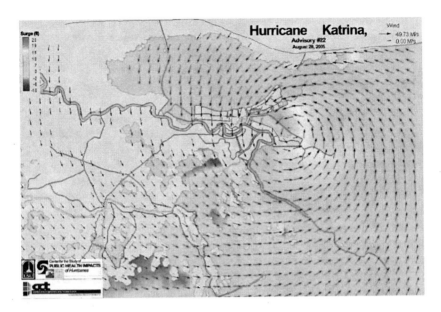

Source: ILIT (2006).

Figure 4.1   Calculated storm surge level when the eye of the storm passed
              the east of New Orleans at about 8:30 am (CDT), 29 August
              2005

east of New Orleans at about 8.30 a.m., 29 August 2005. A calculated
maximum storm surge from the ADCIRC model is shown in Figure 4.2.

The storm surge produced significant overtopping of storm levees
along the lower Mississippi River reaches in the Plaquemines Parish area.
Fortunately, even though the region was almost entirely under water, the
Plaquemines Parish protected corridor is only sparsely populated and the
evacuation of the inhabitants in this area was fairly complete before
arrival of large storm surge. Most breaches were the result of overtopping
and erosion. The breaches occurred mainly in the 'storm' levees while the
'river' levees withstood the storm surge (and waves) without much
erosion.

The storm surge also overtopped the levees in St Bernard Parish, including the overtopping of the Michoud levee. The main levee failures in the
whole system included the Mississippi River Gulf Outlet channel
(MRGO), the Gulf Intracoastal Waterway (GIWW) and the Inner Harbor
Navigational Canal (IHNC).

The massive overtopping flow that passed through the MRGO
frontage levees also overtopped secondary levees built to a lower level of

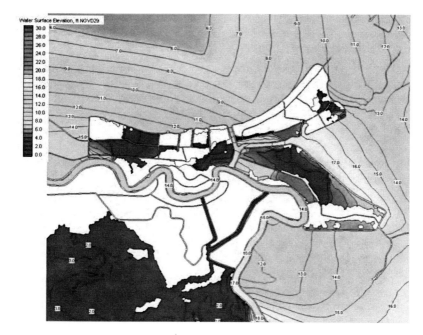

Source: IPET (2006).

Figure 4.2   Map showing calculated aggregate maximum storm surge levels

protection in St Bernard Parish. The overtopping caused an unexpected and rapid flooding of St Bernard Parish, with many homes floating off their foundations.

The storm surge from Lake Borgne overtopped and eroded the levees along MRGO frontage levees and rushed westward over the southeastern corner of the New Orleans East protected section. This was the source of the catastrophic flooding that subsequently made its way across the swamp-lands into the populated areas of New Orleans East.

The same storm surge from Lake Borgne also passed westward into the focused funnel as the water mass entered the shared GIWW/MRGO channel which separates the St Bernard and New Orleans East protected areas. This elevated surge of water then passed through the T-juncture with the IHNC channel. The surging water divided to the north and south along the IHNC channel, overtopping the levees and floodwalls east and west of the IHNC.

As the hurricane progressed northward, the counterclockwise direction of the storm set up a surge toward the south shore of Lake Pontchartrain.

The storm surge sloshed the lake water over a long section of the levees and floodwalls in the New Orleans East protected area causing elevated flood levels within the levee (IPET, 2006).

The 17th Street, New Orleans, and London Avenue canals extend from north of downtown New Orleans to Lake Pontchartrain. The storm surge along the Pontchartrain lakefront did not produce water levels high enough to overtop floodwalls atop these local drainage canal levees. However, a major breach occurred on the 17th Street Canal and two occurred on the London Avenue Canal. These breaches caused catastrophic flooding of the Orleans East Bank protected area. The resulting flooding accounted for more than 40 percent of the total death toll. Contributions to the catastrophic flooding in this area came from the overtopping and breaches along the IHNC channel, but the majority of the floodwater came from the three breaches along the drainage canals.

Extreme flooding caused by levee system breaches, along with communication systems damage caused by the extreme winds, slowed efforts to assess the breaches and determine ways to repair the levees. Some water began to flow out of New Orleans and back into Lake Pontchartrain as the lake resumed its normal water surface elevations of 0.5 to 1 ft above sea level two days after the flooding. The Corps breached other locations to speed the draining process and then plugged the breaches. Sheet piles were driven across the mouths of the 17th Street and London Avenue canals to reduce flow into the effected areas.

ENGINEERING DESIGN LESSONS

Based on the findings contained in the two major reports developed by large investigation teams (IPET and ILIT) in 2006, the major factors for levee failures are listed below:

1. Levee overtopping by storm surge.
2. Floodwall failures due to weak levee soil conditions.
3. Poor levee system maintenance.

The IPET and ILIT investigation teams noted several areas of concern related to improving the engineering design of the hurricane protection system (HPS). Significant changes are needed in the engineering approaches and procedures currently used for many aspects of the HPS. Changes are needed in the design standards, the conceptual approaches considered, the conceptualization and treatment of potential modes of failure, and evaluating performance during design and operation. There is also a need for interactive and independent expert technical oversight and review. Such review would have likely caught and challenged errors and poor judgments that led to failures during Hurricane Katrina (ILIT, 2006, Chapter 15).

An overarching engineering lesson to be learned from the Hurricane Katrina disaster is the importance of life-safety protection systems. The Corps was in the process of constructing a system to provide protection to the New Orleans area. Although Congress had originally mandated the system, over time priorities changed and funding for the system became problematic. Projects were underfunded year after year and design standards were reduced to save money without analyzing the risks created. The costs for recovery and reconstruction significantly overshadow the cost to construct the system. All systems that are designed to protect lives and important infrastructure must require risk analysis before changing design criteria. Engineers, and the populations they serve, should champion the projects in order to focus limited resources on these important systems (ILIT, 2006, Chapter 15).

Floodwall Design and Levee Transitions

The performance of levees and floodwalls varied significantly throughout the New Orleans area. Investigation has shown two main causes for floodwall and levee system breaching: erosion due to overtopping and instability due to soil foundation failure. Of the 284 miles of levee and floodwalls, 169 miles were damaged during the hurricanes.

The performance of levees varied significantly. In some areas, the levees performed well in spite of the overtopping. In other areas, levees were completely washed away after being overtopped. The ability of levee reaches to withstand overtopping without erosion depended on the levee construction material and the severity of the surge and wave action on the levee (IPET, 2006, Vol. V)

The performance of the floodwalls and levee system is divided into two sections related to overtopping and erosion failure, and instability failures. Both sections end with the engineering lessons learned during the HPS failure analysis.

Erosion Failures and Lessons Learned

Breaches due to erosion from overtopping occurred at three locations on the IHNC and at many locations on the MRGO, GIWW and the Mississippi River levees. Approximately 50 miles of earthen levees overtopped but did not breach; approximately 20 miles of earthen levees overtopped with significant breaches; approximately 7 miles of floodwalls overtopped but did not breach; and approximately 2 miles of floodwalls overtopped and were breached. The majority of levees and floodwalls were damaged by overtopping, but did not breach. Hydraulically filled levees, with higher silt and sand content in the embankment material, that were subjected to high overtopping surge and wave action, suffered the most severe damage. Rolled clay levees performed well, even when overtopped (IPET, 2006, Vol. V).

Floodwalls atop the earthen levees consisted of two main types: I-wall and T-wall sections. A single-design cross-section was used for long sections of the I-walls, which was often applied where the protected side ground elevation was lower than considered in the design analysis. I-walls throughout the hurricane protection system were subjected to overtopping and suffered extensive foundation erosion and scour on the protected side of the floodwall. The only exceptions were walls that had paved surfaces adjacent to the walls on the protected side. Significant scour and erosion also occurred at many transitions between concrete structures and earthen levees (IPET, 2006, Vol. 5).

Overtopping of the I-walls led to significant scour and damage in many cases. Overtopping of T-walls did not lead to extensive scour and erosion because the base of the inverted T-wall sections extended over the protected side. Because of the T-wall pile foundations, T-walls are better able to transfer high lateral water loads into stronger underlying foundation materials (IPET, 2006, Vol. 5).

A common problem observed throughout the flood protection system was scour and wash-out at transitions between structural features and earthen levees. The structural features were often higher than the adjacent earthen levee. The difference in elevation resulted in levee scour and washout at the end of the structure. At these locations, the dissimilar geometry concentrated the flows at the levee–structure intersection, causing high flow velocities and turbulence leading to levee soil erosion.

In some cases, the structures were lower than the connecting earthen levees. At these sites, the flow was channeled over the structural feature, causing erosion of soil on the protected side of the structure. Performance in these cases can be improved by providing erosion protection on the inner side of the structures and along the transition section (IPET, 2006, Vol. V).

Floodwall and Levee Transition Lessons Learned

1. Rolled clay fill embankments withstand overtopping without erosion for many hours and should be used to construct levees wherever possible. Armoring can augment existing levee materials to provide improved

erosion resilience (IPET, 2006, Vol. V). The use of highly erodible sand and lightweight shell sand fills should be disallowed in systems protecting major metropolitan regions (ILIT, 2006, Chapter 15).

2. Embedding structural walls within the levee fill, and armoring the transition between levees and floodwalls, can achieve improved resistance against erosion at transitions between earthen levees and structures (IPET, 2006, Vol. V).

3. Design methods should be updated periodically to include the review of recent research and case histories (IPET, 2006, Vol. V).

Levee Stability Failures and Lessons Learned

Four major breaches occurred at one location on the 17th Street Canal, two locations on the London Avenue Canal, and one location on the IHNC due to instability of floodwalls (IPET, 2006, Vol. V).

The four major floodwall breaches resulted from shear failure or erosion and piping through the levee foundation soils at water elevations below the original design level. All foundation-induced breaches had a gap open on the canal side between the levee and floodwall as the water level rose against the wall. Water entering these gaps imposed increased loads on the walls. Where the foundation soil was permeable sand, water flowing down through the gaps increased the water pressure in the sand, reduced the capacity of the foundation to resist loads, and increased the likelihood of erosion and piping (IPET, 2006, Vol. V).

The risks associated with under-seepage flows during 'transient' storm surges were systematically underestimated. This resulted in sheetpile curtains that were extended to inadequate depths at a number of locations, and led directly to a number of the major failures and breaches during Hurricane Katrina (ILIT, 2006, Chapter 15).

17th Street Canal

Field evidence, analyses and physical model tests show that the breach was due to instability caused by shear failure within the clay in the foundation beneath the levee and the I-wall, with a rupture surface that extended laterally beneath the levee, and exited upward through the marsh layer. A key factor in the failure was gap formation. The gap allowed water pressure to act on the wall below the surface of the levee. Another important factor was the low shear strength of the foundation clay beneath the outer parts of the levee and beyond the toe of the levee. A post-Hurricane Katrina analysis, conducted by the ILIT team, clearly indicated that the factor of safety at the I-wall was severely lowered by the water-filled gap. The analysis is presented in Figure 4.3. These two failure mechanisms have significant

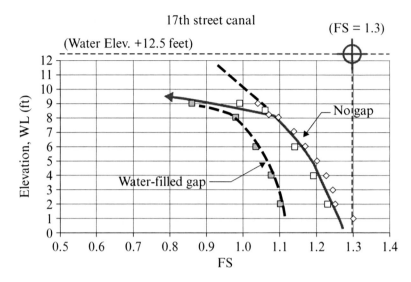

Source: ILIT (2006).

Figure 4.3   Reduction of factor of safety by water-filled gap

system-wide implications. Gap formation and lateral variation of shear strength beneath the levee must be considered for other locations through-out the system when geologic conditions are similar to those at the 17th Street Canal (IPET, 2006, Vol. V).

London Avenue Canal
The south breach on the London Avenue Canal occurred near Mirabeau Avenue. The northern breach was near Robert E. Lee Boulevard. Field evidence, analyses and physical model tests show that the breaches were due to the effects of high water pressures within the sand layer beneath the levee and I-wall, and high water loads on the walls. The London Avenue Canal breaches also had gaps form between the wall and the levee fill on the canal side of the wall. Formation of the gap allowed high water pressures to act on the wall below the surface of the levee and allowed water to flow down through the gap into the underlying sand. High water pressures in the sand uplifted the marsh layer on the landside of the levee, concentrating flow and erosion, removing material and reducing floodwall support.

   Analyses of the south breach showed that erosion is most likely the principal mode of failure, with sliding instability occurring after significant volumes of sand and marsh had been removed by erosion and piping.

Field observations at the north breach indicate that the canal-side levee crest remained intact after the breach, and a playhouse on the property adjacent to the breach was heaved upward as the ground beneath failed. The analyses show that conditions for erosion and piping were present at the north breach, but the more likely cause of the failure was sliding instability. It seems reasonable to assume that the wall on the opposite side of the canal from the north breach, which moved and tilted, must have been close to failure (IPET, 2006, Vol. V).

Inner Harbor Navigational Canal
The IHNC failed in four locations. One of these failures was related to instability. Gaps formed at all breaches on the IHNC, but with the gap, the levee for the north breach on the east side had a safety factor of 1.0 and appears to be the breach that resulted in early flooding of the 9th Ward (IPET, 2006, Vol. V).

Levee Stability Lessons Learned

1. Levee systems should be re-evaluated with regard to the potential modes of failure discovered during the post-Katrina analysis. The evaluation should give appropriate consideration and analysis to underseepage issues including potential embankment instability due to pore pressure induced strength reduction, and potential internal erosion of levee material due to piping (ILIT, 2006, Chapter 15).
2. Design procedures must include the potential failure modes that include formation of the gap on the outboard side of the floodwalls, increasing the lateral forces on the levees (ILIT, 2006, Chapter 15).
3. Many of the I-wall floodwalls are being replaced by the more robust T-wall floodwalls with additional battered piles to help them resist overturning and lateral displacement. T-wall systems will have somewhat increased capacity, but should be analyzed with regard to the 'gapping' failure mode (ILIT, 2006, Chapter 15).

Pump Stations

The information in this section is summarized from Chapter 6 of the IPET report (IPET, 2006). Historically, pump stations have not been considered part of the HPS except where the buildings are a structural part of a levee or floodwall. The pump stations prevent flooding caused by accumulated rainfall and seepage. If the HPS had not failed, the pump stations would have performed as designed to dewater the sub-basins after Katrina had passed.

The flooding from levee overtopping and failure rendered many of the pump stations inaccessible, inoperable or without electrical power. None of the pump station designs protected against local flooding. Katrina significantly damaged one-third of the stations resulting in a 37 percent loss of capacity. Most pump stations require operators and only a few have automatic controls that allow remote operation. Some stations were operational during the storm but were out of service because the operators were evacuated. There are just over 80 pump stations in the four parishes. Some are new and others are approaching 100 years of age. The pump stations vary significantly in their design, construction and capacity. Stations range from large reinforced concrete plants to small-capacity stations housed in metal-frame buildings.

For many stations, power is normally provided by the electrical grid with backup diesel generators or direct-drive diesel engines available when the electrical grid is out of service. Some stations use pumps directly connected to diesel engines, while others utilize 25 Hz power provided by a central generating plant to run the pumps.

Many pumps require priming. The priming process can take as long as 15 minutes per pump. To prevent backflow when the pump is shut down, a siphon discharge valve, a large valve, or gate is installed. If the water level on the discharge side is higher than the invert's high point, reverse flow occurs even with the vacuum breaker valve open. Katrina's storm surge resulted in water levels in Lake Pontchartrain that exceeded the design discharge water levels for some pump stations, resulting in reverse flow.

Reverse flow through unmanned pump stations in Jefferson Parish caused flooding in areas where the HPS did not fail. Reverse flow may have occurred at stations that were later flooded from failed and/or overtopped levees. However, in these instances, the volumes produced by backflow were insignificant in flooding those areas.

The pump station failures in the New Orleans area were investigated. The investigation found four reasons for pump station failure during the storm:

1. Evacuation: all Jefferson and St Bernard Parish pump station operators were evacuated. Only operators at four stations in Plaquemines Parish remained at their stations. Orleans Parish operators stayed at their plants until they were no longer operable. During Katrina, only 16 percent of the total HPS pumping capacity remained operational.
2. Flooding of station equipment: this includes equipment flooded when the levees were overtopped or breached, and pumps that were turned off when they began to only circulate floodwaters through the breaches.
3. Loss of electrical service: failure of primary and backup power supply systems.

4. Loss of lubricating and cooling water: some pump stations rely on potable municipal water services for lubricating and cooling the pumps. Raw canal or flood water is not clean enough for substitution.

Pump Station Lessons Learned

1. Ensure the safety of operating personnel during a hurricane with safe houses at or nearby the pump stations, reinforced control rooms within pump stations, or reinforcing the pump stations themselves to provide protection and creature comforts.
2. Pump stations must withstand hurricane-force winds without significant damage. Modifications are needed to ensure the stations can continue to be operated during peak hurricane-force winds.
3. Pump stations should not be allowed to permit reverse flow (backflow).
4. Trash raking equipment should remain effective during high wind conditions when nobody can be outdoors.
5. There should have been a backup system for the municipal water system at the central generating station for operating its steam turbines.
6. A backup system for the municipal water system should have been provided at stations which use the potable water for lubricating pump bearings.
7. Critical equipment, such as power generators, should have been elevated or protected from 'design' storm water levels.
8. Plant 60 Hz power (for equipment other than pump prime movers) should have remained available when the local utility electrical supply system failed. Generally, this means some plants need small, engine-driven generating sets to be installed for this purpose.

EVACUATION AND TRANSPORTATION LESSONS

Generally speaking, transportation infrastructure is not designed to accommodate evacuation-level demands, as it is not economically or environmentally feasible to build enough capacity to move the city's population away from danger in a matter of hours. Despite these constraints, the evacuation of New Orleans for Hurricane Katrina can be considered as a qualified success. The traffic-count data show that more people were able to leave the city in a short period of time than previously thought possible. The apparent serious failures were in the evacuation of the low-mobility group. Television images of rescue helicopters plucking the stranded people from the rooftops, the New Orleans Convention Center and the Superdome, have overshadowed the success of the highway-based evacuation plan. The number of people that were not, or could not be, evacuated from the city was estimated to be between 100 000 and 300 000 (Wolshon, 2006).

There was significant damage to the Interstate 10 bridge system crossing Lake Pontchartrain, even though the railroad and Highway 11 bridges nearby were relatively undamaged. The Interstate 10 bridge design created structures in the shape of an upside-down bowl with a downward opening face. This allowed air to be trapped beneath the bridge, causing an uplift force due to buoyancy, allowing the bridge deck sections to float off their piers. As the water receded, the decks settled into other locations or fell off the piers. The Highway 11 bridge spans were shorter, allowing less air to be trapped, while the railroad bridge was a solid deck that allowed no air to be trapped (UMR, 2005).

Evacuation and Transportation System Lessons Learned

1. Three measures of major traffic modifications have been found to improve the evacuation significantly:
    *   Staged evacuation plans that identify the order of evacuees starting with the lowest-lying area.
    *   Contraflow loading plans with timelines to maximize capacity and minimize congestion.
    *   Access management plans to help spread the demand to many highways instead of concentrating on the freeway.
2. Plans for evacuating the city's poor population and providing adequate transportation to shelters need to be overhauled.
3. A better communication plan is needed for communicating and evacuating the low-mobility population. This will definitely be the most important issue for all levels of government in establishing and implementing future evacuation plans.
4. Highway bridge decks that cross large water bodies which may be submerged during design events must be designed to prevent buoyancy or be anchored to the piers. If the decks are anchored to the piers, the uplift forces must be considered in pier design as well.

RECONSTRUCTION OF NEW ORLEANS

Since the disaster relief efforts began, reconstruction of the city has been discussed. There are many schools of thought on how the city should be rebuilt to protect the residents adequately. Three options have been discussed for many years. These include: building larger levees and floodwalls; building a city similar to Venice, Italy; and restoring marshlands and moving out of the floodplains. In all of the alternatives, cost-effectiveness needs to be considered and land subsidence needs to be addressed.

The cost to rebuild the Gulf Coast is very high. The money spent by tax-payers nationwide for rebuilding New Orleans must be economically justified. It is not reasonable to ask the entire nation to rebuild an area without providing adequate protection from similar disasters.

Land subsidence is an important consideration in rebuilding. The subsidence is linked to the petroleum industry's extraction of oil and natural gas from the Gulf of Mexico. It is also linked to construction of the Mississippi River levees that prevent sediment deposition by the river in the marshland areas. In order to offer long-term protection to the area, land subsidence in Louisiana must be stopped. If the subsidence continues, protective barriers will begin to lose their protective capacities as soon as they are built.

CONCLUDING REMARKS

The New Orleans HPS is a complex network of levees, drainage channels and pump stations in a very populated area exposed to extreme risks due to its location below sea level. The system is only as strong as the weakest link and so construction and maintenance of the system must always seek to reinforce areas where protection has fallen below the standard of the entire system.

All engineered systems involve risk and there is no way to build infrastructure to protect society against every possible risk. This is due to limited resources, such as money to fund the systems and the time and manpower to design and build the system. For this reason, most engineering projects are built to prevent risks at a level acceptable to society. In the case of New Orleans, the levee system was built to protect the citizens from a fast-moving category 3 hurricane. Hurricane Katrina, a category 4 hurricane, was devastating to the city and its population. The devastation has led society to question the level of protection that should be provided to areas with large urban populations, and what is the most cost-effective approach to reducing the risks.

Since there is no feasible way to provide protection against all risk, warning systems and emergency preparedness must always play a role in the operation and maintenance of life-safety systems. Engineers, politicians and citizens must all be informed and prepared to implement the emergency plan. The emergency plan should include evacuation of the low-mobility population and should be tested before the actual emergency. Training must be provided to prepare emergency responders and decision-makers.

Implementation of the emergency plan helps prevent loss of life and damage to homes and businesses.

All life-safety systems should be periodically reviewed based on updated design procedures and standards. As lessons are learned through research or the failure–investigation cycle, the new information should be reviewed and applied to existing systems. Some of the lessons learned from the failures of the New Orleans HPS include: I-walls do not prevent erosion from overtopping which may lead to breaching; gap formation on the outboard side of levees can exert larger loads on soils and lead to soil failure; construction materials should be evaluated for shear resistance; pumping systems should be designed as part of the HPS; and bridges should be designed for submergence. Funding levels should also reflect the importance of the system. As lessons are learned and applied, life-safety systems and performance will improve throughout the nation, and hopefully, the world.

REFERENCES

Aravamuthan, V. and J.N. Suhayda (1999), 'Real time forecasting of hurricane winds and flooding', Report of Louisiana Water Resources Research Institute, Louisiana State University, Baton Rouge, LA, June.

Bourne, Jr., J.K. (2004), 'Gone with the water', *National Geographic Explorer*, October.

Brouwer, G. (2003), 'The creeping storm', *Civil Engineering Magazine*, June.

Dean, Robert G. (2006), 'New Orleans and the wetlands of Southern Louisiana', *The Bridge,* **36** (1)

ILIT (2006), *Investigation of the performance of the New Orleans flood protection systems in Hurricane Katrina on August 29, 2005*, Draft Final Report, Version 1.2 1 June, Report No. UCB/CCRM- 06/01, University of California, Berkeley, 22 May.

IPET (2006), *Performance evaluation of New Orleans and Southeastern Louisiana hurricane protection system*, Draft final report of the Interagency Performance Evaluation Task Force, June.

NWS – National Weather Service, National Hurricane Center (2006), 'The Saffir–Simpson Hurricane Scale', http://www.nhc.noaa.gov/aboutsshs.shtml.

Suhayda, Joseph N. (1985), *Coastal Flooding Hurricane Storm Surge Model-Volume 1: Methodology*, FEMA Surge Model updated 1 June.

UMR (2005), UMR Public Relations, 'Gravity played role in New Orleans' bridge failures', University of Missouri-Rolla, News@UMR, 28 November, http://news.umr.edu/research/2005/NOLA_bridges_2005.html

Wolshon, Brian (2006), 'Evacuation planning and engineering for Hurricane Katrina', *The Bridge*, **36** (1).

5. Katrina vs. 9/11: how should we optimally protect against both?

Jun Zhuang and Vicki M. Bier

Our report shows that the terrorists analyze defenses. They plan accordingly.
(*The 9/11 Commission Report*, National Commission on Terrorist Attacks Upon
the United States (2004), p. 383)

INTRODUCTION

In the aftermath of the terrorist attacks on 11 September 2001, and
Hurricane Katrina in August 2005, the US government is grappling with
how optimally to protect the country from both terrorism and natural
disasters, subject to limited resources. The all-hazards approach seeks
protections that are effective against all types of emergency events (US
Government Accountability Office, 2005), and was originally proposed to
address this kind of problem. However, this approach does not explicitly
consider the terrorist's analysis and response, and tends to focus more on
emergency response than on prevention.

To our knowledge, only Powell (2005) has formulated a model for allocating defensive investment between terrorism and natural disasters, taking
into account the fact that the level of attacker effort is endogenous – and
Powell's is only an exploratory analysis. In this chapter, we describe a model
for balancing protection from terrorism and natural disasters, and present
results that provide insight into the nature of optimal defensive strategies
in a post-9/11 and post-Katrina world.

The 9-11 Commission Report (National Commission on Terrorist
Attacks Upon the United States, 2004) clearly states that terrorists make
decisions in response to the potential victim's observed strategies. In our
model of this decision process, the attacker and defender are described by
four attributes: (1) the technologies available to the attacker and defender,
represented by the probability of damage from an attack (as a function of
the levels of attacker effort and defensive investment), and the probability
of damage from a natural disaster (as a function of the level of defensive

investment); (2) the attacker's and defender's valuations of the various potential targets; (3) the attacker's and defender's (dis)utilities with regard to the damage caused by an attack; and (4) the attacker's and defender's disutilities for attacker effort and defensive investment, respectively.

Unlike in most past work (for example, Sandler and Arce, 2003; Konrad, 2004; Bier et al., 2007), we represent the level of attacker effort as a continuous variable. This allows us to model the probability of damage from an attack as a function of the levels of both attacker effort and defensive investment. By analogy with the laws of supply and demand governing relations between producers and consumers, we envision attackers and defenders jointly determining their levels of attacker effort and defensive investment, in either a simultaneous or a sequential game. Variables other than the levels of attacker effort and defensive investment (such as the attractiveness of particular targets) are assumed to be exogenous in our model. Farrow (2007) also provides models for attacker and defender optimization problems separately, but fails to link them. For additional applications of game theory in the security context, see for example Major (2002), Woo (2002), Harris (2004) and Bier (2005).

In principle, the defender can deter an attacker in numerous ways – for example, by increasing the opportunity cost of an attack (Frey and Luechinger, 2003), or by making potential targets less attractive (see, for example, Perrow, 2006). However, we focus specifically here on how the defender should allocate defensive investments to reduce the probability of damage from an attack. Similarly, the defender can protect against natural disaster in numerous ways, but we focus on how to allocate defensive investments to reduce the probability of damage from a natural disaster (rather than, say, reducing the impact of that damage through emergency response).

The resulting model for balancing defense against terrorism and natural disasters is described rigorously in Zhuang and Bier (2007). We view this model as a building block toward a more complete understanding of strategic defense against terrorism, and a basis for studying additional types of defenses (such as all-hazards approaches and border security) in future. This chapter provides a relatively non-technical overview of the model and its policy implications.

We begin by describing a basic game between an attacker and a defender. We then compare the simultaneous and sequential formulations of this game, and discuss the defender's first-mover advantage in the sequential game. For ease of exposition, we first present the results of our model for the case of only a single possible target. Even the single-target case allows us to capture the fact that increasing defensive investment can either increase or decrease attacker effort, and to explore the defender's trade-off between protection against terrorism and against natural disasters. We then extend these results to the case of a multi-target system; in particular, we investigate how the existence of multiple targets affects the relative desirability of protection from terrorism versus natural disasters, and explore the effects of risk attitudes on the attacker and defender decisions. Finally, we summarize the policy implications of our work.

ASSUMPTIONS AND PROBLEM FORMULATION

We consider two types of threats: intentional threats (for example, terrorism), and non-intentional threats (for example, natural disasters). The attacker allocates his attack effort optimally among various possible targets. Similarly, the defender allocates her defenses against both terrorism and natural disasters optimally among the targets. For each target, the probability of damage from an intentional threat is a function of both the attacker's effort and the defender's investment in protecting that target against terrorism. By contrast, the probability of damage from a non-intentional threat is a function solely of the defender's investment in protecting the target against natural disasters.

The total attacker utility is the sum of the disutility of the attack effort, and the expected utility of the resulting damage. Similarly, the defender utility is the sum of the expected disutility of damage from both terrorism and natural disasters, plus the disutility of defensive investment against both terrorism and natural disasters.

The probability of damage is assumed to satisfy the following major properties; other, more technical assumptions are discussed in Zhuang and Bier (2007):

1. The probability of damage from an intentional attack equals zero when the attacker effort is zero, and goes to zero as the defender's investment in protection from intentional attack grows arbitrarily large.
2. The probability of damage from an intentional attack is increasing in the level of attacker effort, with decreasing marginal returns.
3. The probabilities of damage from both intentional and non-intentional attacks are decreasing in the level of defensive investment, again with decreasing marginal returns.

As in most applications of game theory, we assume that the attacker and defender have common knowledge about the rules of the game, including the probabilities of damage, the disutilities of attacker effort and defensive investment, the number of targets in the system, and the attacker's and defender's valuations of those targets. We also assume that each party knows the other party wishes to maximize its total expected utility. In other words, the goal of the attacker is to maximize his total expected utility by choosing the level of attacker effort to devote to each target. Similarly, the goal of the defender is to maximize her total expected utility by choosing suitable levels of defensive investment for each target.

SIMULTANEOUS VS. SEQUENTIAL GAMES

We consider both simultaneous and sequential games between the attacker and the defender, as follows. In the simultaneous game, the attacker and the defender decide on the attacker effort and defensive investment simultaneously. (Note that this model can apply even if the attacker and defender do not make their decisions at the same time, as long as neither party knows the other's decision at the time it makes its own decision.) A 'Nash equilibrium' for this game is a pair of attacker and defender strategies such that no player could do better by changing strategies unilaterally.

In the sequential game, the defender chooses a defensive strategy first, and then the attacker chooses his levels of attacker effort after observing the defensive investments. In this case, the optimization problem for the defender is to choose the defensive investment that will maximize her expected total utility, given that the attacker will implement a best response to the defender's investment strategy. The combination of the defender's optimal defensive investment and the attacker's best response constitutes a 'subgame-perfect Nash equilibrium'.

In Zhuang and Bier (2007), we show that the defender's total expected utility in the sequential game is (weakly) greater than the corresponding utility for any possible equilibrium of the simultaneous game (at least if the attacker's best response to the defender's equilibrium strategy is unique). This follows from the fact that the defender always has the option to choose the equilibrium strategy from the simultaneous game even in the sequential game, so some other strategy will be chosen only if it performs at least as well.

Thus, the defender has a first-mover advantage in the sequential game as long as the attacker's best response is unique. Therefore, our model suggests that the defender should in general advertise her defensive investments instead of keeping them secret, in order to use her first-mover advantage. Similar results from attacker–defender models can be found in Bier et al. (2007). The intuition behind this perhaps surprising conclusion is that announcing the defensive investments can help to deter the attacker, or at

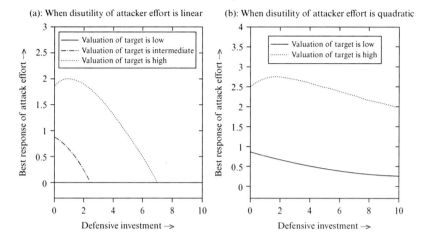

Figure 5.1   Possible attacker best-response functions

least deflect attacks towards less damaging (and less heavily defended) targets.

THE CASE OF A SINGLE TARGET

In this section, we apply our model to a one-target system. In this case, the only two possible outcomes are success or failure of the attack, since the attacker has no choice of targets.

Attacker's Best Response

Figure 5.1 illustrates the attacker's best responses for cases with both linear and quadratic disutility of attacker effort, as follows.

Case 1
If the attacker disutility is linear in attacker effort, there are three possible shapes of attacker best-response functions, as shown in Figure 5.1a. In particular, if the attacker's utility of damage from a successful attack on the target is sufficiently small, then the target will not be attacked at all, regardless of the level of defensive investment, leading to an attacker best response of zero. For a somewhat larger utility of damage, the attacker's best response will be decreasing in defensive investment, reaching zero for large levels of defensive investment. However, if the attacker's utility of

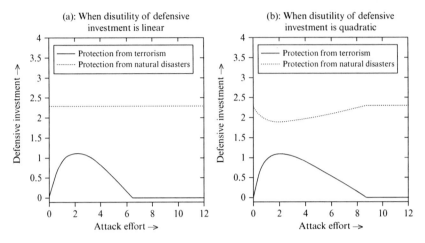

Figure 5.2   Possible defender best-response functions

damage is sufficiently high, then the attacker's best response will be initially increasing in defensive investment, and then decreasing.

Case 2
If the attacker has quadratic disutility of attacker effort (that is, increasing marginal disutility), there are only two possible attacker best responses, as shown in Figure 5.1b. In particular, if the attacker's valuation of the target is low, then the attacker effort will be decreasing in the level of defensive investment, eventually converging to zero. For more valuable targets, the attacker's best response is initially increasing, and then decreasing.

Thus, the attacker's best response eventually converges to zero (if initially positive) as the level of defensive investment grows. However, at low levels of defensive investment, increases in defensive investment can lead the attacker to allocate more effort to attacks (in order partially to compensate for the reduced effectiveness of attacker effort) if the target is sufficiently valuable.

Defender's Best Response in the Simultaneous Game

The defender's best responses are illustrated in Figure 5.2, for the same two cases considered above.

Case 1
If the defender's disutility is linear in defensive investment, then the defender's optimal investment in protection from terrorism is illustrated in

Figure 5.2a. When the defender's disutility of damage is relatively small (that is, the target is not important to her), then the target will not be defended against terrorism at all, regardless of the attacker effort. By contrast, if the defender's disutility of damage is high, then the defender's optimal invest- ment in protection against terrorism will be initially zero, then increasing in attacker effort, and eventually decreasing to zero for high levels of attacker effort. In other words, at low levels of attacker effort, increases in attacker effort will lead the defender to allocate more resources to defense against ter- rorism, in order partially to compensate for the reduced effectiveness of defensive investment. However, at high levels of attacker effort, spending more on defense against terrorism will no longer be cost-effective. Figure 5.2a also shows that the optimal defensive investment in protection from natural disasters is constant in attacker effort for the linear case.

Case 2

If the defender has quadratic disutility of defensive investment, the defender's optimal level of protection from terrorism is similar to that dis- cussed above, as shown in Figure 5.2b. However, the defender's optimal level of protection from natural disaster is now no longer constant, but instead is initially decreasing in attacker effort and then increasing, as shown in Figure 5.2b. Thus, protection from terrorism is a partial substi- tute for protection from natural disaster in this case.

   To summarize, at low levels of attacker effort, increases in attacker effort might lead the defender to allocate more resources to defense against ter- rorism (in order partially to compensate for the reduced effectiveness of defensive investment), but will also generally lead the defender to allocate less to defense against natural disasters. By contrast, at extremely high levels of attacker effort, spending on defense against terrorism will no longer be cost-effective for the defender, so she will spend more on defense against natural disasters.

Equilibrium for the Simultaneous Game

There are three possible types of equilibria for the simultaneous game, as shown in Table 5.1. If the target does not interest the attacker at all, then both the attack effort and the defensive investment in protection against terrorism will be zero. If the target interests the attacker, but the defender's disutility of damage is extremely small, then the attacker effort will be pos- itive, but the defensive investment in protection against terrorism will be zero. Finally, if the target is sufficiently valuable to both the attacker and the defender, then at equilibrium there will be positive allocations of both attacker effort and defensive investment.

Table 5.1 Possible equilibria for the simultaneous game

Case	Attacker effort	Defensive investment
1	0	0
2	+	0
3	+	+

Note that in the simultaneous game, the defender cannot completely eliminate the risk of attack at equilibrium, unless the target is not of interest to the attacker in any case. This changes in the sequential game, where the defender can sometimes completely deter the attacker at optimality, as discussed below.

Equilibrium for the Sequential Game

In the sequential game, we still have the three types of equilibria discussed above, as shown in Table 5.2. In other words, if the attacker's valuation of the target is low, then both the attacker effort and the defensive investment in protection against terrorism will be zero. If the attacker's valuation of the target is high but the defender's valuation is low, then the defender may leave the target undefended even though the attacker effort will be positive. If the attacker's valuation of the target is high and the defender's valuation is moderately high, then both attacker effort and defensive investment will be positive.

However, it is also possible for the attacker to be completely deterred in the sequential game. This will happen when the defender's valuation of the target is sufficiently high relative to the attacker's valuation. This finding reinforces the previous observation that the defender is better off playing a sequential rather than a simultaneous game. (Note, however, that it will not always be optimal for the defender to deter the attacker completely, if the cost of doing so is too high.)

Table 5.2 Possible equilibria for the sequential game

Case	Attacker effort	Defensive investment	Notes
1	0	0	
2	+	0	Same as in simultaneous game
3	+	+	
4	0	+	Unique to sequential game

THE CASE OF MULTIPLE TARGETS

In this section, we discuss games with more than one potential target. We first consider the case where both the attacker and the defender are risk neutral with respect to the level of damage from an attack, and have linear disutility of attacker effort or defensive investment. Then, we explore the effects of risk attitude and convex disutility of effort (that is, increasing marginal disutility) on the equilibria.

Note that in the case of multiple targets, we cannot guarantee that a pure-strategy equilibrium exists, nor that the defender can always constrain the attacker's choices in a sequential game. This is because the attacker and defender best responses may not be unique. However, the existence of a unique best response requires only that the objective function have a unique global optimum; the existence of multiple local optima does not in and of itself cause a problem. Therefore, there are likely to be pure-strategy equilibria even in many multiple-target games (especially when the targets are not homogeneous).

If both the defender and the attacker are risk neutral with respect to damage and have linear disutilities of attacker effort and defensive investment, Zhuang and Bier (2007) show that the multiple-target game reduces to a set of independent single-target games, so the results for a single-target game apply. However, this depends critically on the assumption that the disutilities of both attacker effort and defensive investment are linear (for example, that neither the attacker nor the defender has a budget constraint, so that both the attacker and the defender can allocate as much effort as desired to one target without having to reduce their allocations to other targets).

Effects of Risk Attitude

Perhaps more realistically, we propose that attackers and defenders may reasonably be modeled as risk seeking and risk averse over damage levels, respectively. Although we have not been able to find equilibria for the general n-target case when the attacker and defender are no longer risk neutral, the results for a two-target game (Zhuang and Bier, 2007) yield intriguing insights.

In particular, in a two-target game, Zhuang and Bier (2007) show that, if neither target individually would merit positive attacker effort, then neither risk-neutral nor risk-averse attackers would find it worthwhile to attack both targets, but risk-seeking attackers may do so. Similarly, if neither target individually would merit positive defensive investment in protection from terrorism (or natural disaster), then neither risk-neutral

Table 5.3   Defensive investment for one- and two-target games

	Target 1		Target 2		Total
	Terrorism	Natural disasters	Terrorism	Natural disasters	
For 1 target	1	0	–	–	1.0
For 2 targets	0↓	0.4↑	0.5	0	0.9↓

nor risk-seeking defenders will find it worthwhile to defend both targets from terrorism (or natural disaster), but risk-averse defenders may do so.

The result for risk-seeking attackers, in particular, may help to explain why the terrorists involved in the 9/11 tragedy chose to attack four targets simultaneously. Similar examples include: the four attacks on the London public transport system on 7 July 2005; the four attempted attacks on the London transport system on 21 July 2005; the ten commuter-train explosions in Madrid on 11 March 2004; and the long series of suicide bombings in both Iraq and Israel. In some of these cases, any one target by itself may conceivably not have been sufficiently attractive to be worth attacking, but the prospect of being able to cause larger amounts of damage by attacking multiple targets could have been sufficient to motivate the attackers. Similarly, a risk-averse defender may optimally defend more targets than a risk-neutral or risk-seeking defender would (possibly even including some targets for which the defensive cost is greater than the expected loss due to an attack on that target alone), because of the effect of risk aversion.

Convex Disutility of Effort

Zhuang and Bier (2007) show that if the disutility of defensive investment is linear, then the optimal level of defensive investment in protection from non-intentional threats at equilibrium will be independent of both the attacker effort and the defensive investment in protection from intentional threats. However, this is not true when the disutility of effort is convex (as seems likely to be the case in practice), as illustrated below.

Consider a two-target game with convex disutilities of total attacker effort and total defensive investment, and allow both the attacker and the defender to be risk neutral with respect to the level of damage from an attack. The bottom row of Table 5.3 shows the defensive investment at optimality for this case. Note by comparison with the top row (for the single-target case) that the existence of the second target causes the defender to switch from optimally protecting target 1 against terrorism to protecting it only against natural disaster at optimality. The total defensive investment also falls from 1 in the single-target game to $0.4 + 0.5 = 0.9$ in the two-target game.

The above example indicates that even if protection from terrorism is more cost-effective than protection from natural disasters for a single target, this may no longer be true when additional targets are considered. This is because the terrorist can now redirect his effort among the possible targets in response to the defender's investments. This drastically reduces the defender's ability to allocate her investments in protection from terrorism to those targets that are most cost-effective to defend (Bier et al., 2005), if those are not also the most attractive targets to the attacker. In particular, this phenomenon can reduce or eliminate the desirability of protecting less attractive targets (such as relatively small cities) from terrorism.

SUMMARY AND CONCLUSIONS

In the single-target case, our results indicate that increased defensive investment can lead the attacker either to increase his level of effort (to help compensate for the reduced probability of damage from an attack), or to decrease his level of effort (because attacking is less profitable at high levels of defensive investment). This can either reduce or increase the effectiveness of investments in protection from intentional attack, and will therefore in general affect the defender's optimal allocation of resources between protection from intentional attacks and from natural disasters. In particular, in cases where increased defensive investment causes the attacker to redouble his efforts (due to the attractiveness of the target), defensive investment against terrorism will not be as cost-effective as the defender might have expected based on an exogenous model of attacker effort.

While our model does not formally consider repeated attacks, the 1993 and 2001 attacks on the World Trade Center do seem to represent an example of terrorists 'redoubling their efforts' in attacks against a particularly attractive target. A similar phenomenon can be observed in aviation security, where attackers have resorted to creative attack strategies (from box cutters, to shoe bombs, to liquid bombs) in response to enhanced security, rather than being deterred.

The assumption of endogenous attacker effort in the model described here is critical to capturing important insights into the nature of equilibrium defensive strategies. Therefore, our results emphasize the importance of intelligence in counterterrorism to anticipate not only the attacker's choice of targets, but also likely attacker responses to defensive investments (that is, deterrence vs. redoubling of effort).

In addition, protection from terrorism will tend to become less cost-effective for the defender as the number of targets grows, due to the ability of the attacker to redirect his attack effort to less-defended targets. Thus, even a target that would have been worth protecting in a single-target game may no longer be worth defending from terrorism in a multi-target game. This will in general tend to reduce the effectiveness of protecting large numbers of targets against intentional attacks, and increase the relative desirability of protection from natural disasters (and of all-hazards approaches). This suggests, for example, that the strong emphasis on terrorism defense over natural disaster preparedness at the US Department of Homeland Security may have been misplaced, especially in light of the fact that all but one of the most costly disasters since 1970 was of natural origin (Michel-Kerjan, 2006). However, our results do not call into question the cost-effectiveness of overarching measures such as intelligence or border security; only the effectiveness of attempting to harden large numbers of targets, especially targets of only modest value.

REFERENCES

Bier, V.M. (2005), 'Game-theoretic and reliability methods in counter-terrorism and security', in A. Wilson, N. Limnios, S. Keller-McNulty and Y. Armijo (eds), *Mathematical and Statistical Methods in Reliability, Series on Quality, Reliability and Engineering Statistics*, Singapore: World Scientific, pp. 17–28.

Bier, V.M., A. Nagaraj and V. Abhichandani (2005), 'Protection of simple series and parallel systems with components of different values', *Reliability Engineering and System Safety*, **87** (3), 315–23.

Bier, V.M., S. Oliveros and L. Samuelson (2007), 'Choosing what to protect', *Journal of Public Economic Theory*, **9** (4), 563–87.

Farrow, S. (2007), 'The economics of homeland security expenditures: foundational expected cost-effectiveness approaches', *Contemporary Economic Policy*, **25** (1), 14–26.

Frey, B.S. and S. Luechinger (2003), 'How to fight terrorism: alternatives to deterrence', *Defence and Peace Economics*, **14** (4), 237–49.

Harris, B. (2004), 'Mathematical methods in combating terrorism', *Risk Analysis*, **24** (4), 985–8.

Konrad, K.A. (2004), 'The investment problem in terrorism', *Economica*, **71** (283), 449–59.

Major, J. (2002), 'Advanced techniques for modeling terrorism risk', *Journal of Risk Finance*, **4** (1), 15–24.

Michel-Kerjan, E.O. (2006), 'When insurance meets national security: the unnoticed paradox of critical infrastructure protection', Working paper, University of Pennsylvania.

National Commission on Terrorist Attacks Upon the United States (2004), *The 9-11 Commission Report: Final Report of the National Commission on Terrorist Attacks upon the United States*, Washington, DC: W.W. Norton & Company.

Perrow, C. (2006), 'Shrink the targets continued', *Spectrum*, **43** (9), 46–9.
Powell, R. (2005), 'Defending against terrorist attacks with limited resources', Working paper, University of California, Berkeley.
Sandler, T. and D.G. Arce M. (2003), 'Terrorism and game theory', *Simulation and Gaming*, **34**, 319–37.
US Government Accountability Office (2005), 'Homeland security: DHS's efforts to enhance first responders' all-hazards capabilities continue to evolve', GAO-05-652.
Woo, G. (2002), 'Quantitative terrorism risk assessment', *Journal of Risk Finance* **4** (1), 7–14.
Zhuang, J. and V.M. Bier (2007), 'Balancing terrorism and natural disasters: defensive strategy with endogenous attacker effort', *Operations Research,* **55** (5), 976–91.

6. Worst-case thinking and official failure in Katrina

Lee Clarke

The event we call 'Katrina' created two disasters, one in New Orleans and one on the Gulf Coast. In this chapter I will focus on the New Orleans disaster, for several reasons. The two disasters are of different characters. The one on the Gulf Coast was certainly devastating. I do not downplay the significance of the damage done there or the terrible suffering of the people from all those places that are, as victims said, 'gone'. However, the Gulf Coast has been destroyed before. Hurricane Camille, in 1968, was stronger than Katrina, a category 5 storm, but was in many ways similar.[1] That storm also caused billions of dollars in property damage. Pielke et al. (1999) note that: 'the storm was called the greatest catastrophe ever to strike the United States and perhaps the most significant economic weather event in the world's history'. Exactly the same superlatives were used for Katrina. There were other similarities. My main point in comparing Camille with Katrina is that the Gulf Coast, or more accurately large parts of it, has been destroyed before. This means that there are mental models for how to rebuild it.

That is not the case in New Orleans. There are problems there for which we have little or no precedent. Perhaps the most daunting of problems is that we have no idea how to rebuild a culture. There are good reasons to doubt that this can be done. Culture exists in people and their artifacts (and in how people relate to their artifacts) and both are substantially 'gone' from New Orleans. Regarding people, the categories that are most relevant are the kinds of individuals, and the range of ethnic, racial and religious groups that used to be there. It was in those peoples' minds and relationships that the history and culture of New Orleans lived. Much of the physical infrastructure, too, held not just bodies but families and communities. There was a lot of old and elegant architecture in New Orleans that is now gone. But I mean something more than beautiful buildings that people used and loved. I mean that the sense of place-attachment in New Orleans was partly a product of how people lived and thought about their physical surroundings. Even if there were some commitment on the part of leaders and organizations to rebuild the rich and complicated culture that used to exist in New Orleans, there is no expert knowledge that I am aware of that we could draw upon for guidance. The social infrastructure, necessary to sustain the culture, was washed away with the physical infrastructure.

Compared to rebuilding culture, rebuilding New Orleans physically would be relatively easy. This would entail massive investment – in the levees and in building structures so that they could withstand high winds and high waters – but in principle it could be done. But the commitment is not there. There is no program from the White House that is predicated on an imperative of making New Orleans safe. I doubt that any Congressperson's election or re-election campaign will lament the loss of New Orleans as much as it will lament the loss of family values and the 'threat' of homosexuals marrying. There is no social movement animated by the drowning of New Orleans. It is not going to happen. And, it must be said, there is a good argument that New Orleans is better off with a smaller footprint. Perrow (2007) argues that the most important thing we can do to make ourselves safer is to reduce the size of targets (whether the targets be buildings, size of chemical plans, complex systems, or people living in dangerous places). Since the levees will not be built to withstand a category 5 hurricane, and since people are not required to rebuild at heights, say, over 8 ft, and since another large hurricane will eventually revisit New Orleans, a smaller footprint – drastically reduced population levels – makes sense as an investment in future safety. Still, the loss will be great.

It is customary to say that the event we refer to as Katrina happened on 29 August 2005. But that is a convention. We could as easily say that the disaster, or catastrophe as Quarantelli (2006) has called it, began when the decision was made to build levees that were not strong enough. Even that would be a mystification because of course there was no single decision to build insufficiently, or perhaps poorly. Surely no one in the US Army Corps of Engineers said, 'Let's do inadequate work.' And there was no single decision at all. A detailed history would surely show a slow accretion of choices, no one of which was sufficient to doom New Orleans. Another way to tell the 'Katrina' story would be to set the beginning of the disaster with the failures, at all levels of government, to develop adequate evacuation plans, or to allow building in low-lying places.

Just as the beginning of the Katrina story is difficult to pin down, so too is its end. No one who has walked with the ghosts among the piles of rubble that used to be homes and businesses would say the disaster ended when the water was pumped out of the city. I would not say the disaster is over. The disaster we know as Katrina continues in the loss and trauma that its victims endure.

Thinking about the boundaries of Katrina, or any disaster, as malleable poses unconventional challenges for prediction, and for thinking about responding effectively to calamity.

KATRINA, THE WORST CASE

In *Worst Cases* (Clarke, 2006) I identified the fundamentally cultural nature of the definition of a 'worst case'. There is no objective measure of a worst case, or at least no interesting objective measure. When planes crash we count the bodies and then use that count to decide whether it was the worst crash ever. But even then we differentiate. The worst single-aircraft accident? The worst commercial accident? The worst domestic accident? And so on.

According to the conventional measure – body count – Katrina was far from a worst case. That title, inside the United States, is still held by the 1900 hurricane that devastated Galveston. It is known as the Galveston Hurricane because hurricanes did not then have names. It is estimated that the Galveston Hurricane killed at least 8000 people. And if we look beyond US borders Katrina looks like a mosquito bite, rather than a worst case. For there have been several cyclones in Bangladesh that have killed 200 000 people or more.

If Katrina was not a worst case in the simple sense of body counts, it certainly was in a more complex, cultural accounting. The basic feature of a worst case is that it overwhelms the imagination. Before Katrina it was unimaginable that so many people would lose their lives in a hurricane in the United States. I should add that it was not unimaginable to everyone. There were a great many scientists from different walks of academic and professional life who imagined Katrina all too clearly. None of those scientists, however, were in positions of power that would have enabled them to act on their knowledge.

There were several ways that Katrina overwhelmed imaginations and thus qualifies as a worst case. We saw considerable outrage throughout the US because of the apparent abandonment of poor, black people. That the outrage seems short-lived does not gainsay that it was at one time real. Americans are just not used to seeing that level of suffering, because it was so clearly preventable.

Katrina was a worst case in a broader sense. It is astounding to realize the range of social position and demographic category from which people from and around New Orleans say, 'We never thought it could be this bad.' Even scholars whose specialty is disaster or environment (the study of disaster is a subset of environmental studies) were caught off guard. These are not people who have an interest in denial, as we might say of some officials in the Federal Emergency Management Agency (FEMA). And these certainly are not unintelligent people. Far from it. But the extent of the inundation, and the great depth of loss, overwhelmed the imaginations of so many. That is why it is a worst case to them.

If Katrina had been imagined, perhaps there would have been greater preparations, perhaps more people would have got out of the way. Had the worst-case possibilities been sufficiently appreciated, the likelihood that New Orleans would be allowed to drown would have been diminished. Perhaps some official or officials, or some organization or organizations, would have been assigned or even assumed the responsibility for thinking through the implications of having a metropolitan area in America die. Recall Perrow's argument (Perrow, 2007) that New Orleans' footprint should be smaller. Even if that is so, Katrina was the wrong way to make it so and realizing – even publicizing – the worst-case possibilities of topped or breached levees could have provided powerful rhetoric in making the case for deliberate shrinkage.

One way to have more fully realized the potential disaster is to engage in possibilistic thinking. In *Worst Cases* I argue that a probabilistic to risks, or any action, is but one way to engage in rational thought. Probabilism is extremely important, let me emphasize. It is correct to say that the likelihood of safely traversing the country is enhanced by taking a jet airliner rather than driving (the threat of explosive liquids notwithstanding). But probabilism is not the only way to reason about the risk of flying. If your plane is 40 000 feet in the sky and both engines stop running, or a wing falls off, or you run into another airplane, talking about probabilities of living or dying is irrational. That is why people become nervous when their plane bounces around like a child on a trampoline. The pilot says after landing, 'Welcome to Los Angeles. You've just completed the safest part of your trip.' And she's right. The passengers applaud because they have survived what to them was very nearly certain death. And they are right too.

From a probabilistic point of view worrying about the danger from a nuclear power plant makes no sense. Utility executives or the Nuclear Regulatory Commission uses probabilistic arguments to dismiss the fears of local protest groups (say, near Indian Point which is about 25 miles north of Manhattan). The protest groups understand that the likelihood of a meltdown is low. What worries them are the consequences if a nuclear plant should have a particularly bad day.

A problem is how to foster and even institutionalize possibilistic thinking. Of course possibilistic thinking can be used cynically to frighten people to support policies or programs they would not otherwise support. This issue is clear enough. We need more systematic research on the conditions that give rise to possibilistic thinking, and how best to fold that into disaster policies.

PREDICTION AND RESPONSE

Katrina poses profound if somewhat obvious challenges for disaster planning. The most obvious of these challenges is that if officials and organizations could not plan effectively for Katrina, what hope can we have that they will plan effectively for something worse? For here was a case in which all the outcomes were foreseen and, from a certain point of view, could have been prevented. Researchers at the University of New Orleans had conducted surveys of vulnerable people and so, in principle at least, they could have been identified beforehand and their evacuation ensured. It was also known that the city was subsiding, that the wetlands around New Orleans were disappearing, and that the levees were pointedly not designed to withstand a category 5 hurricane. It was known that New Orleans was a 'bowl' that could easily fill with water.

So why was new Orleans left to drown?

1.   Bureaucratic bungling.
2.   Jurisdictional uncertainty.
3.   It couldn't be prevented.
4.   It was in someone's interest.

INSTITUTIONAL DISASTERS

Because the hazard, which is to say the source of the threat, was a natural event, Katrina is often called a natural disaster. That is a mistake that shifts attention away from all the human actions that made the tragedy possible and even likely. Indeed, the term 'natural disaster' is an instance in which common sense categories are adopted in scholarly discourse with insufficient critique. Scholars have for years tried to flesh out the distinction between technological and natural disasters. Most of their efforts have been focused on the differences in meaning that attend disaster types. These labors have borne fruit, especially regarding the way that legal conflicts and corporate power create psychological and social damage in communities.

But I propose that it is time to abandon the natural–technological distinction. There is no such thing as a natural disaster. Using that phrase is a convention that no longer leads to productive scholarship. There is, after all, no disaster without people. Had no one been in the path of Katrina there would have been no disaster. This, and not the hazard or source of danger, should be our first principle of analysis.

Rather than organizing our thoughts in terms of an old-fashioned distinction, perhaps we should speak of different kinds of 'institutional disasters'. Institutions put people in harm's way, and they intensify the damage that a hazard poses. Institutions fail to protect people. Institutions cause disasters. In New Orleans, the complex of institutions that contributed to the disaster includes the following:

- Poverty. Not only poor people suffered. But the likelihood of dying was positively correlated with being poor. Micro and macro factors were involved. An example of the former is the cognitive frames used by people to make sense of the threats they faced. Some of those who stayed, and did not die, said they 'did evacuate'. They just didn't evacuate far enough. They evacuated the way most people evacuate – to the homes of family or friends. The problem was that their family's house was only a few blocks away. An example of a macro factor is that poor people do not command sufficient political power that would require someone to protect their interests. The comparison with the disaster response whenever the Berkeley Hills catch fire seems apt.

- Structure of work. The unemployment rate in the New Orleans metropolitan area pre-Katrina was 14.8 percent, one of the highest in the nation. Unemployment is obviously related to poverty. The point here, though, is that over time New Orleans saw the decline of blue-collar, unionized labor and the rise of service work that came with the entertainment industry. Such work does not facilitate the accumulation of wealth nor provide the sense of a secure future.

- Political indifference. Not only is it the case that those who were left behind were mainly poor and black. They were also outside the mainstream of political life. Put differently, the political and economic system did not take care of those people before the levees broke, and it did not take care of them afterward.

- Residential patterns. More than half of the American population now lives near the water. In Florida over 80 percent of the population lives within 20 miles of the coasts. We are moving to dangerous places and New Orleans was more dangerous than most.

- Insurance. The Lower Ninth Ward had a strange set of characteristics. While 36 percent of the Ninth's population was poor (largely minority), 59 percent of the ward owned their own homes (many of which had been passed down through generations). But without a mortgage there would be no requirement to have flood insurance.

Thus did home ownership – usually thought of as a guarantor of safety – increase people's vulnerabilities.
- Environmental. Destruction of the watershed.
- Diminution of FEMA over time; loss of political support; dilution of mission.

In proposing the idea of institutional disaster as a replacement for the 'natural' or 'technological' labels I mean to help move debates away from merely responding to calamity. Our political and economic system will probably always be more reactive than proactive concerning extreme events. But it is possible to overemphasize the aftermath, rather like measuring the tiretracks after the truck has run over someone. In *Worst Cases* I argue that disaster is as normal as love, joy, corruption and crime. My insistence on this point is so that we move away, intellectually and politically, from seeing disaster as special, thus requiring special theories. We obviously need policies aimed at muting the worst effects of disaster: evacuation plans, relocation assistance, and so on. But if we really want to lessen the crushing financial and human toll of disasters – both of which are going to get worse – our efforts should be directed at the institutional make-up of society. The concept of institutional disaster leads us in that direction.

FAILURES OF RESPONSE

Every level of government, and most levels of private enterprise, failed in New Orleans. We saw planning failures, as noted. We also saw, as Tierney (2007) has said, elite panic. Elite, or powerful, panic (Clarke et al., 2006) is an over-reaction to a potential threat. Social science writings on disaster have for years documented the rarity of panic in disaster. What has been neglected is panic among those in positions of power or authority for example, mayhem reports. Elsewhere, my colleague Caron Chess and I have suggested that what elites are most likely to panic about is the possibility of public panic (Clarke et al., 2006).

Another kind of failure in Katrina was what Perrow (2007) analyzes as 'executive failure'. Executive failure is different from being overwhelmed, or underinformed, or even making a mistake. He says that executive failure is 'where deliberate, knowing choices are made by top executives that do harm to the organization and/or its customers and environment'. It is the willful turning away from situations, facts and the like so that short-term interests are maximized while long-term responsibilities (for example, safety) are forsaken.

THE COUNTER-CASE: US COAST GUARD

Failure was not universal. The US Coast Guard performed admirably. If this performance was a fluke, the 'lesson learned' is depressing indeed. For it means that the next time the only reason we will see success of any sort is chance. But if there are structural sources of the Coast Guard's success then there is some solace, because that would mean it is possible to design offices and organizations for more timely and effective response.

THE MOST PESSIMISTIC CONCLUSIONS POSSIBLE

Worst-case, or possibilistic, thinking emphasizes consequences and outcomes. It is not a replacement for but a complement to probabilistic thinking. As a form of rationality, probabilistic thinking does not serve us well in evaluating risks when the likelihood of occurrence is extremely small. Since that is where, nearly by definition, extreme events happen, we need another way to talk about remote and dangerous futures. Although not inherent, a possibilistic approach definitely has its pessimistic side. In that vein, and in the spirit of sparking discussion, I will draw the most pessimistic conclusions I can think of.

There has been a great deal of criticism leveled at the Department of Homeland Security (DHS) for the Katrina failures. The criticism of ineffective response is earned and just. The mandate of the DHS is to protect us, and not just from terrorists; disaster preparation and response is part of the mandate. Still, that is a blunt judgment. Others, more incisively, account for the FEMA's failure by pointing out that folding the FEMA into a very large bureaucracy took away its independence. But that, too, is insufficient. The Coast Guard is part of the DHS too.

The Coast Guard story is important, but it is exceptional. As has happened in other disasters, for instance the collapse of the Twin Towers, the public acted well and officials and organizations failed. Let us then conclude that we cannot count on organizations to be effective. It is worse than that. Let us assume at the outset that executive failure will be the norm, as will be organizational ineptitude. From Katrina – am I generalizing from a sample of 1? But then, how many failures does it take? – we can conclude that the environment-scanning functions of DHS offices and organizations do not work. They did not sufficiently warn and did not sufficiently respond. Imagine a Boeing 777 loaded with al Qaeda operatives takes off from Saudi Arabia. Central Intelligence Agency (CIA) officials on the ground there have known for some time that the flight was planned, and they notified the White House when the plane took off. The CIA recommends a shoot-down.

But the order to shoot does not come because someone, not the Attorney General, said that shooting down a civilian airliner was illegal and politically costly. The plane lands and the terrorists disperse to lead a coordinated attack, the details of which are classified. That was Katrina.

NOTE

1. http://sciencepolicy.colorado.edu/about_us/meet_us/roger_pielke/camille/report.html; incorp citation.

REFERENCES

Clarke, L. (2006), *Worst Cases: Terror and Catastrophe in the Popular Imagination*, Chicago: University of Chicago Press.
Clarke, L., C. Chess, R. Holmes and K.M. O'Neill (2006), 'Speaking with one voice: risk communication lessons from the US anthrax attacks', *Journal of Contingencies and Risk Management*, **14** (3), 160–69.
Pielke, R.A., C.W. Landsea, R.T. Musulin and M. Downton (1999), 'Evaluation of catastrophe models using a normalized historical record: why it is needed and how to do it', *Journal of Risk and Insurance*, **18**, 177–94.
Perrow, C. (2007), *The Next Catastrophe: Reducing our Vulnerabilities to Natural, Industrial and Terrorist Disasters*, Princeton, NJ: Princeton University Press.
Quarantelli, E.L. (2006), 'Catastrophes are different from disasters: some implication for crisis planning and managing drawn from Katrina', *Social Science Research Council Paper*, New York.
Tierney, K.J. (2007), 'From the margins to the mainstream? Disaster research at the crossroads', *Annual Review of Sociology*, **35**, 503–25.

7. Risk, preparation, evacuation and rescue

Edd Hauser, Sherry M. Elmes and Nicholas J. Swartz

INTRODUCTION

Recent, current, and future threats to populations and the built environment around the world, whether the threats are by natural elements or by human-induced threats or actions, represent complex strategic and tactical planning challenges. They also present challenges to decision-making processes that must be understood by local, state and federal emergency managers, as well as by individuals and families who are at risk from approaching or ongoing disasters. One basic decision, if a choice is possible, is whether and when to evacuate in the event of a pending or ongoing disaster.

This chapter explores the risks and other factors that were considered by New Orleans residents as Hurricane Katrina approached, beginning with the preliminary efforts of area residents and visitors to prepare for the storm. Included are data that describe the assessments by individuals as decisions were being made on whether or not to evacuate. We also briefly touch on the transportation needs of the disadvantaged in mass evacuations.

Considerable attention over the years has been given to research and development of modified trip generation, trip distribution, and assignment models of projected trip volumes, transportation modes and routes. Hurricane evacuation modeling is a current focus of significant efforts to adapt these traditional models, particularly in southeastern and Gulf states.

Currently used models do not possess the capability of considering fact-based estimates of the individual choices made by families and individuals. For example: do we evacuate or not? Do we have a plan? How do we get to a shelter? If evacuating, do we know where we are going? Do we have a staging area to meet family or friends after evacuating? How can we get there? By what means? Do we leave when the weather service is just encouraging evacuation, or do we wait until a mandatory evacuation is announced? Do we know the level of congestion on evacuation routes? These are some of the considerations that are a part of the risk assessment process in these situations, by both public officials and individual citizens.

To be valid, decision support tools must account for the dynamic and adaptive interactions of the threat (in this case, what became a category 5 hurricane) and the populations at risk (specifically for this treatise – the City of New Orleans). Individual-based computational simulation modeling (also called agent-based modeling) holds considerable promise to provide a framework for simulating and analyzing the non-linear macroscopic impacts of the numerous microscopic interactions within the population at risk, in both anticipated and unanticipated scenarios. In New Orleans, there was a definite cascading effect of one disaster immediately following the other (that is, the breached levees and flood led to loss of power and communication systems, isolated and stranded individuals, significant numbers of fatalities, and so on).

The potential for loss of life, serious injury, and the physical, economic and psychological aspects of catastrophes need to be estimated *ex ante* in order to develop a database that can be used in simulations, as well as helping managers make mitigation decisions. As stated above, agent-based or individual-based simulation modeling will help researchers and practitioners to be better prepared through the use of such models in training, exercises and ultimately in real-time response to disasters.

This chapter does not go into details of such simulation models, however. It is about developing a database that describes real-time individual choices made by the residents of New Orleans leading up to the morning of Sunday, 28 August 2005. The findings of this study will be useful in validating such simulation models.

As an initial 'lesson learned' (or at least, a lesson documented), one conclusion of this research, as well as other studies conducted in the aftermath of Katrina, is that individuals need to be better educated about the risks, and better informed in advance about specific imminent risks.

The National Science Foundation (NSF)-supported research discussed in this chapter and undertaken by a team from the University of North Carolina at Charlotte, Duke University and Tulane University aims to understand better how decisions are made about whether to evacuate or not to evacuate. The authors gratefully acknowledge the support of the NSF in supporting its 'quick response' grants to researchers in the aftermath of Katrina. We also acknowledge the excellent collaborative effort by our statistician research colleagues: Dr David Banks of Duke University, Dr John Lefante of Tulane University, and his graduate assistant Maria Sirois.

In our research, 291 residents or former residents of New Orleans and Orleans Parish were interviewed about their evacuation decisions prior to Katrina's landfall on Monday, 29 August. Interviews were conducted during the months of February through June 2006. Subsequently, the following metrics are being examined as part of the study of the database:

- Geographical space (zip code location, elevation of property, relationship to the levees, and so on).
- Socio-economic factors (age, race, housing tenure, income, family members at home, previous experience with hurricanes, and so on).
- Mobility characteristics (lack of personal transportation, disabilities, institutional residency, number and mode of relocations after rescue, and so on).

The results of this study will be used to develop a risk profile to help local authorities decide whether to designate and enforce a mandatory evacuation, how best to publicize that decision for the best result, and the optimal time to do it. Not all of the metrics included in the study are related in this chapter. Those factors not included herein will be given further attention in the ongoing analysis of a very rich database that is being studied by the combined research team.

RISK ASSESSMENT: THE FALLACY OF PREVIOUS EXPERIENCE

The 'science' of risk analysis has been well documented in the literature, and one of the better definitions was described by Lee Clarke in his 1999 book *Mission Improbable: Using Fantasy Documents to Tame Disaster.* Clarke's point is well taken in that:

> . . . the more planning, the less uncertainty. This is so because more planning means decision-makers have collected more information and made more benefit–cost deliberations than is the case when they do not plan. (Clarke, 1999, p. 10)

It is common knowledge, however, that time, lack of focused effort, perhaps 'selective memory', and even chance, conspire against organizational readiness. It is our contention that this applies to individuals and families as well, and one can therefore be burdened with the onerous task of, as Clarke writes:

> . . . effectively transforming uncertainty into risk even without a sufficient experiential base or conceptual scheme appropriate for interpreting history. This problem is especially critical for [individuals] whose failures may entail catastrophe. (Clarke, 1999, p. 12)

It is evident from our survey of 291 residents or former residents of New Orleans that a majority of the population relied only on past experience rather than conducting a rudimentary cost–benefit analysis that took into account the size and intensity of the approaching storm. A sample of responses to the first question, 'What did you expect to happen, based on reports and [your] previous experiences?', revealed that most respondents simply did not think the storm would be 'that bad'. A few of the comments made by respondents reflect the majority of residents' risk assessment prior to the storm making landfall:

- 'just wind damage';
- 'thought it would be minor like previous hurricanes';
- 'the hurricane will just miss us';
- 'expected the hurricane to veer to the east';
- 'didn't expect too much to happen';
- 'expected it to be high tides and flooding';
- 'nothing like that happened before'.

Interestingly, not one respondent to the survey mentioned that they expected the levees to fail. Several mentioned flooding from the rain, and only two individuals interviewed mentioned an expectation that the levees might be topped, but no one responding to the survey mentioned the word 'breached' or 'fail' in their responses.

We do not wish to overdo our reliance on Clarke's conventional wisdom, but his description of how organizations and individuals calculate risk is also informative:

> Risk becomes normalized through simile, which means uncertainty-to-risk transformations are fundamentally rhetorical transformations . . . Behind the rhetoric are experts, organizations, and organized publics who must bear the brunt of decisions made in their name. So it is important not to forget that uncertainty-to-risk transformations have histories of social relations, themselves embedded in clashes of vision and conflicts of interest. (Clarke, 1999, p. 101)

These same social relations, visions and conflicts affect individuals as well as organizations. Taking these into account, all individuals in high levels of responsibility, such as emergency management managers, mayors, governors, cabinet secretaries, and their advisors, all need to have some understanding of what motivates people to understand and assimilate risk better as they make individual decisions. This applies to both private sector organizations and to government.

Gary Klein might argue that a quick decision, based on instinct or 'gut feeling', is a better way to make decisions in such situations. This notion has some merit that we should consider. His emphasis on using intuition rather than rational decision-making is well taken in most workplaces, and in an emergency situation, quick thinking is undoubtedly needed:

> Sometimes, [an individual] can get caught up in the details of quantifying the likelihood of each risk. Often, risk analyses are designed to figure out how much of a safety margin is needed in order to achieve safe operations . . . and the team may be better served by appreciating the limitations of the plan. (Klein, 2003, p. 101)

We should mention that Klein's book, although written in the post-9/11 era, does not necessarily push this method of decision-making for emergency managers or individuals facing a disaster. In reality, however, decision-makers, whether private citizens or public officials, often make their decisions in such situations based on intuition gained from past experiences rather than on an assessment of the facts surrounding the current event. Klein's work is instructive in understanding that approach to decision-making.

STUDY METHODOLOGY

Figure 7.1 shows the study area boundaries of the city of New Orleans. Randomly selected coordinates are shown, although interviews were not conducted at each selected location. There were three separate target groups of interviewees that made up the survey sample of residents and former residents of New Orleans and Orleans Parish. The number of interviews completed for each of the three groups was: personal interview group – 113, computer-automated telephone interview (CATI) telephone survey of residents with land lines – 109, and CATI telephone survey of residents who were reached on their personal cell phones – 69.

Personal Interview Group

Respondents were interviewed by a team of six researchers from Tulane University and UNC Charlotte during the period of February through June 2006. Locations were selected as randomly generated geographic coordinates, with the number of potential interview points being proportionate to the density of housing in each 'cell' of the grid shown in Figure 7.1. The process for identifying locations to attempt to conduct personal interviews was aided by using a global positioning system (GPS) receiver to locate an intersection of two streets closest to each randomly selected geographic coordinates. Interviewers would then locate the nearest dwelling (house, apartment or FEMA trailer) which housed an available,

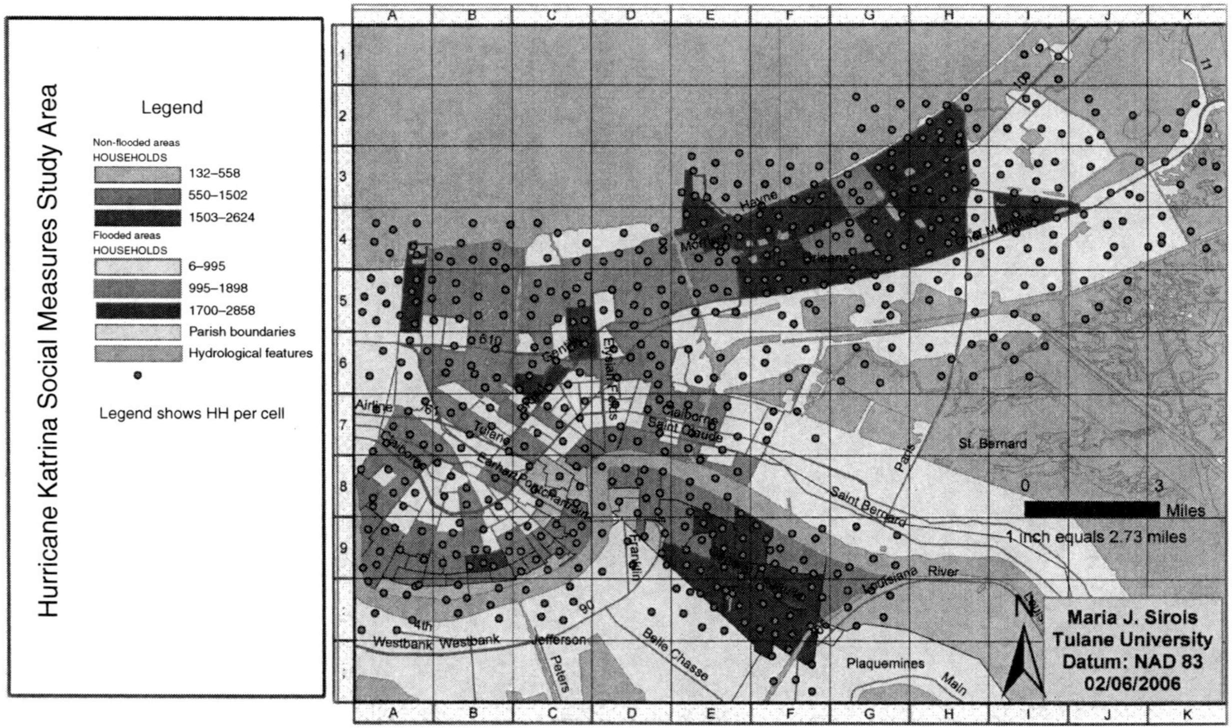

Figure 7.1   Study area, flood depth and data points

eligible respondent. In some cases, the nearest occupied dwelling was more than a block from the intersection. Up to three attempts were made to contact people in each dwelling that was identified. In other cases, there were no occupied houses within several blocks or within sight of the designated intersection.

At least one personal interview was conducted in 32 of the 42 cells that cover the land area within the city boundaries (see Figure 7.1). Cells consisted of land areas of approximately 7 square miles.

A higher proportion of personal interviews were completed in those cells that had some dwellings occupied or were being restored during the February through June survey period. Those cells in which a lower proportion of interviews were completed were in cells that were largely uninhabited (such as the Lower Ninth Ward, East New Orleans, Lakeview and others). The personal interviews, although important to add in-depth, spontaneous dialogue and insights into the disaster, were very difficult to locate, set up an interview, and conduct an interview due to the severe disruption of peoples' lives in these areas.

Most of the respondents in this group were working on their homes during the day with generator power, and spending their nights either in a trailer on-site, or traveling to temporary housing in surrounding parishes as far away as Baton Rouge, due to safety and security issues in the communities around their homes. Figure 7.2 shows the concentration of interviews conducted in each zip code, with the darker areas showing locations of more interviews. There were ten zip code areas that resulted in 11 to 30 completed interviews.

Land Line Telephone Group

Respondents in this group were interviewed using a computer-automated telephone interview (CATI) system, using randomly generated numbers of New Orleans land line telephone exchanges. Those individuals who were contacted by interviewers from UNC Charlotte in June 2006, in general, were residents living in areas not flooded by the hurricane rains or by the breach of the levees.

These residences are in elevated land areas of the 'bowl' that is commonly used to describe New Orleans' topography. Notable among such areas is an approximately ten-block wide ribbon of land along the Mississippi River called the 'Garden District' (see Figure 7.2). Like the downtown area and the French Quarter, these riverside residential areas were the first to have electricity and telephone service restored following the hurricane, and residents were generally not displaced by the flooding, at least for extended periods of time.

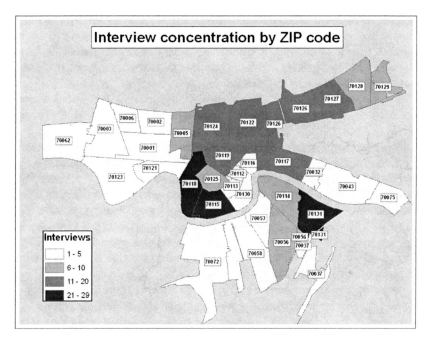

Figure 7.2   Interview concentration by zip code

Cell Phone Group

These respondents were also interviewed using the UNC Charlotte CATI system, after developing randomly generated numbers in three-digit cell phone exchanges known to include the city of New Orleans. These exchanges were compiled from personal contacts within the city, since they were not available from the cell phone providers. Many of these respondents were former residents of New Orleans who had been displaced by the storm and were living in other areas (Baton Rouge and Alexandria, Louisiana; Houston, Texas and elsewhere) at the time of the interviews in July 2006. This group had the highest proportion of persons rescued from their homes, the Superdome, the Convention Center, and other locations.

Response rates for these two telephone surveys (completed interviews per personal contact with a qualified interviewee) were 35 percent for the land line group and 38 percent for the cell phone group.

Again, persons interviewed in all three groups were randomly selected. The analysis shown in this chapter relates responses for the three groups combined. The basic analysis tabulates responses categorized by age and race as dependent variables.

The spatial distribution of the three groups (those contacted by land line and cell phone interviews, and personal interviews) are identified by zip code in Figure 7.2. A general assessment of the location of interviews completed can be compared with the maximum depth of the floodwaters by a visual scan of Figures 7.1 and 7.2. Figure 7.1 is coded with maximum depth of floodwaters caused by breeched levees, up to 28 ft in some places. In those areas indicating no flooding took place, in effect, isolated minor flooding did affect other parts of the metro areas during and immediately after the storm passed over New Orleans on 29–31 August.

PERCEIVED RISK BY AGE AND RACE

With little likelihood that many individuals are so oriented to a systematic review of realistic risks and alternative courses of action during an actual crisis, the quick decisions that are often necessary in such situations utilize perceived risk as a surrogate for the actual risk. In the case of Katrina, even the perceived risk was greatly underestimated by New Orleans residents since the survey data suggest that very few factored in the risk associated with a breach in the levees. The numbers of respondents, all of whom were heads of households or living alone, for each group of the responding population, are as follows. Of the 291 respondents, 52 were under 35 years old, 126 were in the 35–54 age group, 61 between 55 and 64, and 52 older than 64. Also, 156 of the respondents were white and almost all the remainder were African-American.

The four age groups chosen for this analysis were based on 'stages of life' characteristics. For example, individuals and heads of households (HH) in their twenties and early thirties are likely to have no children or smaller children; those aged 35 to 54 are more likely to have children at home; those aged 55 to 64 are more likely to be individuals or couples in 'empty nest' situations; and those over 64 are more likely to be living alone or in group homes or institutional settings.

Note that throughout the remainder of this chapter, all graphics except as noted will be developed for age and race as dependent variables. The percentages shown on the vertical bar for each subgroup relates the percentage that responded affirmatively to each question. Therefore, in Figure 7.3, using as an example the number of respondents in the less than 35 age group, 36 out of 52, or 69 percent, had previous experience with hurricanes. Overall, the total number of respondents who answered in the affirmative totaled 220, or 75 percent of the total sample of 291.

One interesting aspect of this particular metric is that there is a high degree of uniformity among respondents of different races and in different age groups. A logical and expected result of this analysis is that those in the

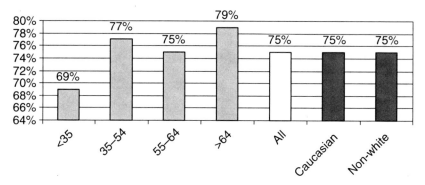

Figure 7.3   Previous experience with hurricanes

younger age group were less likely to have had direct experience with previous hurricanes.

Two factors were more likely to have been taken into account by respondents in advance of Katrina coming ashore in Louisiana: (1) the residents'
previous experience with hurricanes; and (2) their understanding (or lack
of understanding) of the elevation of their residence compared with sea
level (Figure 7.4), or their location in a potential flood zone (Figure 7.5).

Relying on the factor of previous experience, it would appear that the
majority of residents were perhaps lulled into a more relaxed approach
to preparedness and especially evacuation. This was due to the majority
(75 percent) of residents or previous residents having experienced one or
more hurricanes. Hurricane Ivan was the storm most often cited, having hit
the Louisiana coast in the summer of 2004. The comments made by respondents indicated that although Ivan was a major hurricane there were few
fatalities attributed to the storm, and there was little flooding, therefore the
impact of Katrina would probably not be any worse. They were wrong.

Interesting anecdotal information collected indicates that some residents
recalled riding out, or evacuating, up to 10 or 12 storms during their life on
the Gulf Coast. Up until Monday, 29 August 2005, residents were collectively lulled into a false sense of security.

Figures 7.4 and 7.5 relate the understanding of owners or renters prior
to Katrina concerning the relationship of their home's elevation to the sea
level and flood plain designation. The more telling metric in this analysis,
however, was the fact that large numbers of residents did not know the elevation of their dwelling, and nor did they know whether it was in a potential flood zone (Figures 7.5 and 7.6).

For the entire group of 291 respondents, 47 percent (137 respondents)
understood that their residence was either lower than sea level, located in

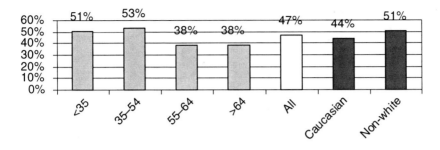

Figure 7.4 Residence below sea level

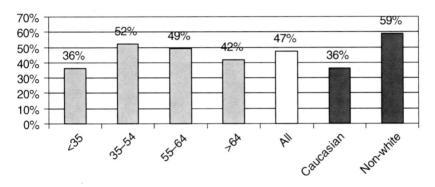

Figure 7.5 Home located in flood zone

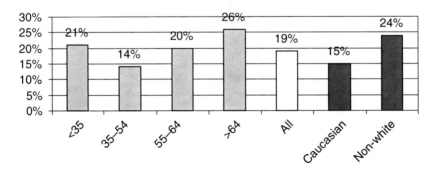

Figure 7.6 Did not know elevation pre-hurricane

a flood zone, or both; 34 percent (99 respondents) understood that their residence was not in a flood zone or below sea level; and 19 percent did not know if either condition existed (55 respondents, approximately one in five).

Relating these data to the subgroups characterized by method of conducting the interviews, 15 percent of the telephone group, 16 percent of the cell phone group, and 45 percent of the personal interviewees did not know the answer to the question about their home's elevation or location. Possible explanations for the significantly higher percentage in the group of respondents involved in personal interviews was that those interviewed in this group were on the average quite a bit older than the other two group averages.

Over 50 percent of those in the personal interview group were over 60 years old, whereas the proportion of the other two groups over 60 years old was less than 30 percent. Interestingly, there was a greater difference between Caucasian and non-white respondents' understanding of the flood zone issue than there was between these two groups in their understanding of their residence location relative to sea level.

PRE-HURRICANE PREPARATION BY AGE AND RACE

For those residents who did understand that Hurricane Katrina was a considerably violent storm, there were preparations made to support the actions being taken, relevant to either evacuating or preparing to ride it out. Both those deciding to evacuate and those staying had taken precautions such as boarding up their residence and removing loose yard furniture. Pets were obviously a consideration. However, the three most common preparatory actions taken by all sub-groups were: (1) boarding up their residence; (2) moving personal belongings to a higher elevation in their residence or outbuildings; and (3) buying food and water.

Interviewees were given an open-ended question concerning their preparation before the hurricane hit the city, but then they were furnished a list of preparatory actions that many people take when faced with a storm that will potentially force them from their homes. A partial list of the most common actions, and the proportion in each age group that took those actions, is shown in Table 7.1. Major differences in pre-hurricane actions among respondents were that the older group were less inclined to fill their vehicles, purchase extra food or water, or board up their homes. This may be attributed to one of two factors. First, the fact that older people lived in a higher (and therefore potentially drier) elevation of the city, and therefore did not think that a flood, should it occur,

Table 7.1   Preparatory actions by age group

	<35 (%)	35–54 (%)	55–64 (%)	>64 (%)
Elevated belongings	15	53	19	13
Bought food and water	21	43	21	15
Bought fuel for vehicle	17	43	21	19
Bought generator	44	4	5	16
Bought extra gas	11	37	26	26
Boarded up their residence	12	49	26	13
Withdrew cash from bank	19	45	19	13
Took no action	18	44	14	24

would threaten their residence. The second explanation for decisions made by this group is that although they were not making preparations to leave, there was a lower vehicle ownership rate among these residents and there were fewer vehicles that needed to be filled up. A third possible explanation is that there would be some in this group that may not have been physically able to carry out any of these tasks if they had no one available to help them.

The age group that made the most preparatory actions were those in the 35 to 54 age group. Their responses to this question tended to be in the 40 to 50 percent range whereas the other groups were in the 10 to 30 percent range.

The most startling and unexpected statistic to be observed among the respondents was that over 20 percent overall made absolutely no effort to prepare for Katrina.

EVACUATION BY AGE AND RACE

To understand better the actions of residents of New Orleans before 29 August 2005, it should be pointed out that Hurricane Katrina was one of those storms that did not move in a manner that was predicted by the models run by the National Hurricane Center. After crossing the southern part of Florida on Thursday, 25 August as a category 1 storm, Katrina covered more of a geographical area than most classic hurricanes, but did become intense very quickly. In spite of only being a category 2 storm on Friday, 26 August, Louisiana's Governor declared a state of emergency for the entire state. Not too many people in New Orleans appeared to be impressed, and even fewer were motivated to evacuate on Thursday or Friday.

Over the course of those two days the storm became a category 4 (maximum wind speeds of 155 mph) and at 4 p.m. on Saturday the 27th,

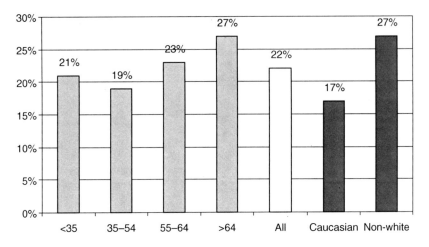

Figure 7.7   Evacuated before 28 August 2005

the Governor ordered the reversible lanes to be open on Interstates 10, 12 and 55, effectively nearly doubling the capacity of the freeways north and west. At 5 p.m. on Saturday, 27 August, the Mayor of New Orleans issued a voluntary evacuation notice.

At 8 a.m. on Sunday, 28 August, Katrina was measured as a category 5 storm (wind speed 175 mph). In our study of interview data from New Orleans residents, Figure 7.7 shows that only 22 percent of respondents evacuated prior to 28 August (in other words, before midnight on Saturday, 27 August). A larger proportion of non-whites (27 percent) than Caucasians (17 percent) had evacuated by midnight on Saturday.

One other transportation-related action was taken on Sunday 28 August. At noon the Regional Transit Authority directed that their buses were to be sent to ten shelter locations throughout the city in order to transport people to the Superdome. Apparently the 'on the fly' plan for sheltering shifted to consolidating people in the Superdome rather than scattered in the ten out-lying shelters

At 10 a.m. on Sunday, the New Orleans Mayor issued a mandatory evacuation order. Sunday morning was also the busiest time for residents and visitors staying in New Orleans to try to leave the city. The proportion of residents that attempted to evacuate between midnight on Saturday and noon on Sunday more than doubled. Figure 7.8 indicates that 53 percent (153 individuals or families) were evacuating by noon on 28 August. This number is cumulative and includes those 63 respondents that evacuated before 28 August.

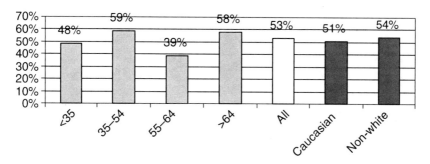

Figure 7.8   Evacuated before noon 28 August 2005

Individuals and families evacuating on Sunday found that even with the use of the reversible lanes on Interstate Routes 10, 12 and 55, it was a nightmare making any headway. The normal two-hour travel time from New Orleans to Baton Rouge was taking ten hours, at an average speed of approximately 12 mph.

It should be noted that a similar situation occurred three weeks later (23 September) when Hurricane Rita skirted the New Orleans area, and poured massive amounts of rainfall on south-western Louisiana and the Texas Gulf Coast and the Houston area. Approximately 3 million people tried to evacuate from the Galveston to Houston urban corridor by going west and north. In the case of Rita, 137 people died in the evacuation. A single vehicle crash involving a bus transporting nursing home residents resulted in a vehicle fire in which 23 elderly people died. Many of the other 114 Rita evacuees died by succumbing to heat exhaustion due to stalled vehicles parked on and beside freeways-turned-parking lots. On the other hand, most of the fatalities in New Orleans were attributable to drowning.

No specific estimate was made of heat-related deaths attributable to Katrina, although there were over 1800 fatalities in the three-state area, and Louisiana accounted for 1580 of them.

By the end of Sunday, more than 80 percent of the city's 484 000 residents had either left or were attempting to leave. Those without transportation were stranded. The Superdome obviously took most of the evacuees, and by 9 p.m. on Sunday night, about 20 000 people were sheltered there. Other timeline events over the next several days are as follows:

- Monday, 29 August, 6 a.m. CDT – Katrina makes landfall near Buras, Louisiana as a category 4 storm with winds of 145 mph.

- Monday, 8:00 am – storm surge sends water over the Industrial Canal, and a barge breaks loose and crashes through the floodwall, opening the first breach.
- Monday, late morning – 17th Street Canal levee is breached.
- Monday, noon – electrical and phone service fails, including cell service.
- Monday, 4:00 pm – Two levees on the London Avenue Canal fail.
- Tuesday, 30 August, late afternoon – the USDHS declares Katrina an Incident of National Significance, and a coordinated federal reponse is ordered.
- Tuesday evening – evacuees at the Superdome reach nearly 25 000.
- Wednesday, 31 August, 10 am – FEMA announces that a convoy of buses will take evacuees from the Superdome to the Astrodome in Houston.
- Wednesday evening – evacuees at the Convention Center number about 3000.
- Thursday, 1 September – up to 20 000 people crowd the Convention Center.
- Thursday evening – the first busload of evacuees arrives in Houston. From there, many were scattered to destinations throughout the southern part of the US, many being moved several times – up to eight different destinations for some of the respondents to our surveys.
- Saturday, 3 September – American Airlines dispatches three aircraft to evacuate people from Louis Armstrong International Airport to Lackland Air Force Base in San Antonio. Over 50 missions were flown by several air carriers over the next five days.
- Sunday, 4 September – cruise ships are dispatched to the Gulf Coast to serve as temporary housing.
- Monday, 5 September – New Orleans Mayor issues a second mandatory evacuation order.
- Friday, 23 September – Hurricane Rita hits the southwestern Louisiana and the Galveston–Houston areas; some of the levees in New Orleans are breached, and some of the wards, particularly the Lower Ninth Ward, are flooded again.

Many respondents in the group that were isolated in the tragedy at the Superdome and/or Convention Center also indicated that their inability to leave was due to the absence of a vehicle in the household (Figure 7.9). Several interviewees in this group, particularly those living on a bus route or trolley route, indicated that they did not own a personal motorized vehicle. Some, however, were able to evacuate by catching a ride with some other driver not a member of the same household.

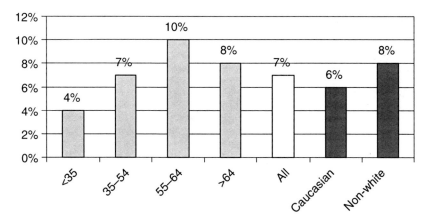

Figure 7.9   Those that could not leave: no transportation

Some respondents indicated they could not leave their residence, or had walked to either the Superdome or Convention Center sites. Therefore, there was a total of 270 individuals that made a choice of whether to leave or ride out the storm.

The next three figures (Figures 7.10–7.12) illustrate the percentage in each subgroup of the population that revealed the principal factors in their decision to leave. Some people had more than one reason to evacuate; however, the three most prevalent reasons to leave were in response to an evacuation order (5.5 percent), TV or radio coverage (30 percent), and personal or family decisions (29 percent). Family considerations obviously included children in the household, infants, elderly relatives or friends, and of course pets.

The largest difference among these data showed that only 19 percent of Caucasian respondents based their evacuation decisions on personal decisions or family considerations, whereas 41 percent of non-white respondents based their decision on this factor.

One additional transportation-related factor that was included in the interviews was the mode of transport used in evacuations. Figures 7.13 and 7.14 illustrate the proportion of each population subgroup that evacuated using personally owned vehicles (POVs) or rode with others.

For those respondents in the personal interview sub-group, a greater proportion indicated evacuation by alternate modes of transport – approximately 10 percent, whereas only 46 percent evacuated in their personal or family vehicle.

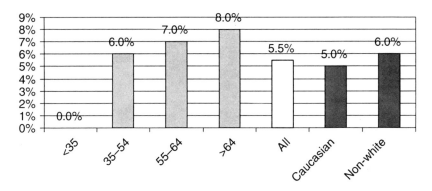

Figure 7.10 Key factor on decision to leave: direct order

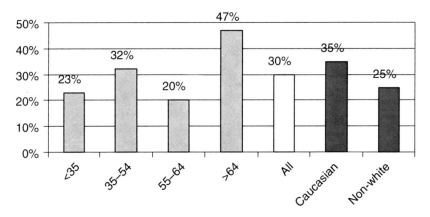

Figure 7.11 Key factor on decision to leave: TV coverage

RESCUE OF RESIDENTS AND VISITORS

Other than individual, volunteer efforts of rescuing individuals stranded on rooftops or in trees, in buildings, and so on, there was little formal rescue activity during the course of the day on Monday, 29 August. On that day, as related previously, the levees on the canals connected with Lake Pontchartrain failed, and the significant level of flooding in much of the city began. On Tuesday, 30 August, the day the floodwaters reached their highest elevation, more organized rescue efforts were launched, and this effort continued for two or three weeks as the flood waters subsided.

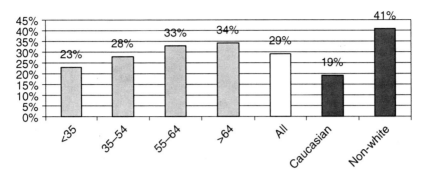

Figure 7.12   Key factor on decision to leave: family decision

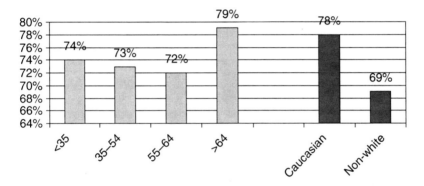

Figure 7.13   Method of evacuation: personal vehicle

While approximately 80 percent of New Orleans's population evacuated before or during the storm, approximately 10 percent of the population had to be rescued by some other means, according to public information provided by the city and various agencies involved. People were rescued by family, friends, neighbors, other volunteers or organized emergency responders. As in most crises and disasters, the prevalent 'first responders' were those individuals who carried out in some cases extraordinary 'Good Samaritan' deeds.

Figure 7.15 illustrates the data on those people rescued at some point in time after the storm had passed. Overall, agencies involved in the rescue after 29 August estimated that somewhere between 7 percent and 10 percent of residents (34 000 to 48 000) were rescued at some point in time. Approximately one-third of respondents who were rescued sometime after 29 August were rescued by boat, either motorized or non-motorized.

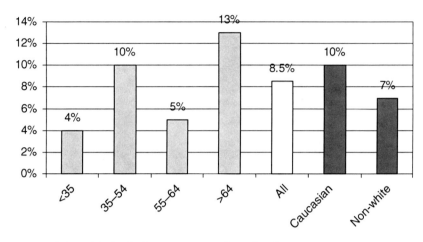

Figure 7.14   Method of evacuation: rode with others

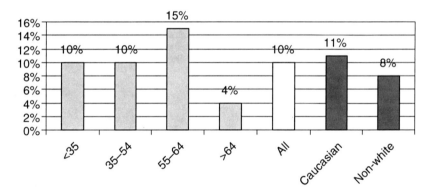

Figure 7.15   Rescued after 28 August 2005

Another third were rescued and evacuated from the city by bus, and the other modes used in rescues included helicopters and other military vehicles. After the initial rescue and transfer to various collection points, the mode was by bus or by air to transfer people to other parishes and cities in Louisiana, Texas, and elsewhere.

It was reported by the US Coast Guard that overall more than 12 500 individuals were rescued by air sources, including the Coast Guard, Navy, Air National Guard, commercial and general aviation aircraft, and others (Baron, 2006, p. 150). This number would include those rescued by heli-

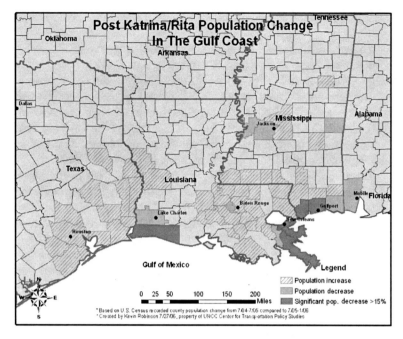

Figure 7.16   Population shifts following Katrina and Rita

copters in the flooded areas and fixed-wing aircraft transporting evacuees from area airports, including Louis Armstrong International.

POPULATION CHANGES

The US Bureau of the Census, as well as the states of Alabama, Louisiana, Mississippi and Texas, have tracked changes in population on a county-by-county basis since the two hurricanes, Katrina and Rita. The data shown in Table 7.2 and Figure 7.16 are estimated population shifts between 1 July 2005 and 1 January 2006. The parishes and counties that experienced most of the population shifts in terms of percentage of pre-Katrina population are dramatically shown in Figure 7.16. This figure shows both decreases in population and corresponding increases. Table 7.2 shows the Census data on population shifts: Orleans Parish decreased by 280 000 people in the six months between 1 July 2005 and 1 January 2006.

LESSONS DOCUMENTED

Researchers dealing with applied research studies – those that are efforts to effectuate positive and productive change in the way our country or state or community carries out its business, both in the public and private sectors – usually want our reports and speeches and research papers to result in 'lessons learned'. If the results of this chapter or any other documentation on Hurricane Katrina are not translated into simple, easily understood 'fact sheets' and used to change the way that our nation and communities address hazards and disasters, then we have not done our job.

Therefore, this concluding section is in terms of 'lessons documented'. The learning will take at least a few more years and perhaps (but hopefully not) decades. The sad, unfortunate reality is that we, Americans in general, will predictably wait until a disaster strikes near home and the 'disaster de jour' generates interest in finding out what went wrong this time.

Based on the on-site work of a team of six researchers in New Orleans, and an equal number of telephone interviewers, we can identify several lessons from our database and approximately 1000 contact hours with Katrina victims that hopefully will reach at least one responsible public official in a hurricane-prone part of the country. Observations or 'lessons documented' are as follows.

The 'Level of Knowledge'

Loosely paraphrasing a now-famous phrase that relates to another international crisis, a certain US Secretary of Defense declared at a news conference

Table 7.2   Population change by parish or county

County/Parish	Population on 1 July 2005	Population on 1 January 2006	% change
Mobile, AL	393 585	391 251	−0.6
Jefferson, LA	448 578	411 305	−8.3
Cameron, LA	9 493	7 532	−20.7
Orleans, LA	437 186	158 353	−63.8
Plaquemines, LA	28 282	20 164	−28.7
St Bernard, LA	64 576	3 361	−94.8
Hancock, MS	46 240	35 129	−24.0
Harrison, MS	186 530	155 817	−16.5
Jackson, MS	134 249	126 311	−5.9
East Baton Rouge, LA	396 735	413 700	+4.3
Harris, TX (Houston)	3 647 656	3 740 480	+2.5

concerning the insurgency in Iraq that 'we know some things we don't know, but there are some things we don't know that we don't know'. There is actually a great deal of wisdom in this statement, and it is especially disturbing when we hear emergency response officials in exercises or training sessions or reports to public officials responding to inquiries about disaster preparation details by stating: 'That's been taken care of already'.

Such thinking and attitudes are likely to result in not having any understanding as to those things that we don't know that we don't know about. Those 'things' constitute the absence of a knowledge base that often results in greater losses at the time of a disaster. The lesson is: we (all Americans) need to expect the unexpected and be as prepared as possible, at a reasonable level, for situations that we have not experienced before. Unfortunately, as our research in New Orleans has revealed, a vast majority of residents rejected the notion that they would be wise to expect the unexpected.

Worst-Case Scenarios

In order to be prepared for such 'worst-case' disasters, we need to recognize that we will never be completely, totally prepared. Disasters will happen and people will be injured, killed or subjected to gravely infectious diseases. Lee Clarke has made a case for 'pre-emptive resilience', the type of response to disaster that we learn about in the story about the passengers on United Airlines Flight 93 on 9/11. Dr Clarke states:

> The power of individuals and their networks shouldn't be underestimated . . . It should be bolstered and facilitated by organized government agencies. Rather than centralize, we ought to shore up resilience at the levels of society where people actually live, where they experience disasters . . . (Clarke, 2003, p. 173)

Personal and Family Evacuation Plans

As pointed out above, about one-fifth of our respondents did nothing to prepare for Katrina at the time and a much greater proportion had no disaster plan in place. The lessons documented by the authors about Katrina, based on our research, indicate that at a minimum, each family or group of any description should have a plan that is clearly understood and preserved by each individual in the group.

Minimum requirements should include: alternative evacuation routes; means of transportation; final destination (at a safe distance from the projected impact zone of the storm, with a specific meeting time, day, and place); means of communication; and specific plans for children and pets.

These factors are dramatized in the case of Katrina because as of mid-May 2006, there were still over 300 Louisiana residents listed as 'missing' (FEMA News Release, 18 May 2006).

The Cascading Effect

Infrastructure systems are interconnected, with each dependent on others for seamless operation. There is a growing science in identifying through geospatial techniques the way that information systems, various utilities, transportation networks, financial networks, public policy-making and decision-making networks, plus public and private managers of critical infrastructure, and so on, are all connected. For example, a loss in electrical power may result in lack of pumping power for natural gas or other fuel that is needed in a cold-weather emergency.

In New Orleans, the cascade was enormous and continues to increase even into late 2006 and beyond. The cascade started with hurricane force winds and heavy rainfall; it then spread to overtopped and breached levees; floodwaters up to 12 feet; lost electrical power and communication systems; random fires; random looting; hazmat (hazardous materials) spills in the flood waters; inaccessible transportation routes; untreated sewage in the floodwaters; undrinkable water; submerged POL (Petroleum, Oils and Lubricants) tanks floating to the surface; wild animals, snakes, alligators throughout the area; trapped residents; no supplies or electricity or medical services in hospitals; closed hospitals; closed services; former residents not returning; severely diminished economy; and the list goes on – for years after the disaster.

Crisis Communications

We are reminded that the first responders at many incidents, crises and disasters are not trained professionals but individuals who just happen to be at the scene. Providing appropriate information and training opportunities for the public needs to be a continuing mitigation type activity of EMAs (Emergency Management Agencies). Following the 'ramping-up' of the US Department of Homeland Security, a number of communities around the country initiated a 'Citizens Corps' program that prepared neighborhoods, commercial developments, university campuses, businesses and industries to at least have an ongoing awareness of the various types of possible threats and hazards.

Many communities have initiated Emergency Response Councils and other community groups to keep the issue before the public. In New Orleans, where 30 percent of respondents indicated that their decision to leave was based primarily on TV coverage, the value of effective media relations and a wide-area crisis communications plan cannot be overemphasized.

Expect the Unexpected

Emergency management agencies need to take the lead in encouraging communities to think collectively and systematically about issues that have not been thought about before, or perhaps even rethink about issues that have been covered before, but just not lately. New Orleans held an exercise in 2004 called 'Hurricane Pam'. That exercise included a segment of working through a scenario of 'What would we do if the levees fail?'

One year later, the unexpected actually happened. During the year prior to the storm, the levee-breaching scenario should at least have been considered as a possibility and included in at least one full-scale exercise, turning the unexpected into the possible. As political scientist Scott Sagan has observed, 'things that have never happened before happen all the time' (http://understandingkatrina.ssrc.org/Clarke/p 4, 2005).

BIBLIOGRAPHY

Baron, Gerald R. (2006), *Now is Too Late: Survival in an Era of Instant News*, Bellingham, WA: Edens Veil Media.
Bedient, P.B., A. Holder, J.A. Benavides and B.E. Vieux (2003), 'Radar-based flood warning system applied to tropical storm Allison', *Journal of Hydrologic Engineering*, **8** (6), 308–18.
City of Charlotte and University of North Carolina at Charlotte (2004), 'Charlotte Center City Evacuation Plan', Charlotte/Mecklenburg Police Department.
City of Charlotte and University of North Carolina at Charlotte (2006), 'Evacuation protocols and transportation guidelines: Attachment 3 to Annex I, Mecklenburg County All Hazards Plan' (draft), Office of Emergency Management.
Clarke, Lee (1999), *Mission Improbable: Using Fantasy Documents to Tame Disaster*, Chicago: University of Chicago Press.
Clarke, Lee (2003), *Worst Cases: Terror and Catastrophe in the Popular Imagination*, Chicago: University of Chicago Press.
Federal Emergency Management Agency (FEMA) (2004), 'Using HAZUS-MH for risk assessment', FEMA Publication 433.
Federal Emergency Management Agency (FEMA) (2006), 'Travel trailer/mobile home safety for hurricanes, floods, and tornadoes', FEMA Publication TF-016.
Florida Division of Emergency Management (2005), 'Citizen emergency information', www.floridadisaster.org.
Greene, R.W. (2002), *Confronting Catastrophe: A GIS Handbook*, Redlands, CA: ESRI Press.
Harris, E.A., et al, (2005), 'Determining disaster data management needs in a multi-disaster context', Final Report, NSF Grant CMS-0353175, proceedings of a

workshop on Disaster Studies Data Requirements, North Carolina State University, Raleigh. www.disasterstudies.uncc.edu.

Hasenberg, C.S. and F.N. Rad (1999), 'Lessons learned in a level two HAZUS analysis for buildings and lifelines in the Portland, Oregon metropolitan area', Technical Council of Lifeline Earthquake Engineering Monograph.

Hauser, E.W., S.M. Elmes and J.K. Robinson (2006), 'Emergency Evacuation Modeling', Center for Transportation Policy Studies Monograph, University of North Carolina at Charlotte.

http://cc.msnscache.com/cache.aspx?q.

http://en.wikipedia.org/wiki/HurricaneKatrinadeathtollbylocality.

http://mceer.buffalo.edu/publications/Katrina/default.asp, 'Rapid damage mapping for Hurricane Katrina' (2006), Poster session at the Hazards Conference, Boulder, CO.

http://news.galvestondailynews.com/story.lasso?tool=print&ewed=03ccaa2ed8838088, posted February 26, 2006.

http://onlinepubs.trb.org/webmedia/2006am/431Reep/description.htm.

http://onlinepubs.trb.org/webmedia/2006am/278Renne/description.htm.

http://onlinepubs.trb.org/webmedia/2006am/278Sanchez/description.htm.

http://onlinepubs.trb.org/webmedia/2006am/431Walsh/description.htm.

http://understandingkatrina.ssrc.org/Clarke/p4, posted June 28, 2006.

http://www.atsdr.cdc.gov/emergencyresponse/importancedisasterplanning.

http://www.cas.sc.edu/geog/hrl/home.html, 'Current projects at the Institute for Hazards Research at the University of South Carolina', Poster Session at the Hazards Conference, Boulder, CO, August 2006.

http://www.charmeck.org/Departments/Police/About+Us/Departments/Center+City+Evacuation+Plan.htm.

http://www.ecu.edu/hazards/ 'The Center for Natural Hazards Research at East Carolina University', Poster Session at the Hazards Conference, Boulder, CO, August 2006.

http://www.fema.gov/news/newsrelease.fema?id=25341.

http://www.fhwa.dot.gov/reports/hurricanevacuation/, 'Report to Congress on Catastrophic Hurricane Evacuation Plan Evaluation'.

http://www.heritage.org/Research/HomelandDefense/SR06.cfm, 'Empowering America: A Proposal for Enhancing Regional Preparedness'.

http://www.hurricanesafety.org/newpoll4.shtml, 'National Hurricane Survival Initiative'.

http://www.kansas.com/mld/kansas/news/14676795.htm?template=contentModules, posted Friday, 26 May 2006.

http://www.noaanews.noaa.gov/stories 2005/s 2484.htm.

http://www.nola.com/printer/printer.ssf?/base/news-3/114846737260150,xml@coll=1, posted 1 June 2006.

http://www.nytimes.com/2005/12/07/opinion/o7opchart.html, posted 7 December 2005.

http://www.ohsep.louisiana.gov/factsheets/DefinitionOfaHurricane.htm.

http://www.orlandosentinel.com/news/shs-ap-hurricane-survey, posted 20 July 2006.

Klein, Gary (2003), *The Power of Intuition: How to Use Your Gut Feelings to Make Better Decisions at Work*, New York: Doubleday/Random House.

Kovel, J.P. (2000), 'Modeling disaster resource planning', *Journal of Urban Planning and Development*, **126** (1), 26–38.

Levin, A. and P. Eisler (2005), 'Many decisions led to failed levees', *USA TODAY*, 2 November.

Louisiana Recovery Authority Support Foundation (2006), 'Collective strength: research, planning, sustainability, 2006 recovery survey, citizen and civic leader research summary of findings'.

North Carolina Institute for Disaster Studies (2004), 'Institute history and planning for future activities', www.disasterstudies.uncc.edu.

Perry, R.W. (2003), 'Incident management systems in disaster management', *Disaster Prevention and Management*, **12** (5), 405–12.

Pradham, A.R., D.F. Laefer and W.J. Rasdorf (2003), 'Infrastructure information management system framework requirements for disasters', Raleigh, CA: North Carolina State University.

Price, J. (2006), 'Few heed hurricane lessons', *Raleigh News and Observer*, 22 July.

Risk Management Solutions, Inc. (2005), *Hurricane Katrina: Profile of a Super Cat. Lessons and Implications for Catastrophe Risk Management*, report, Newark, CA.

Tierney, K.J., M.K. Lindell and R.W. Perry (2001), *Facing the Unexpected: Disaster Preparedness and Response in the United States*, Washington, DC: National Academy of Sciences, Joseph Henry Press.

Udin, N. and D. Engi (2002), 'Disaster management information system for southwestern Indiana', *Natural Hazards Review*, **3** (1), 19–30.

United Nations Office for the Coordination of Humanitarian Affairs (2004), 'An overview of OCHA's emergency services', http://ochaonline.un.org.

8. Not Katrina: the Thames Barrier decision

Chang-Hee Christine Bae and
Harry W. Richardson

INTRODUCTION

In the aftermath of Katrina, there have been recriminations about the failures of flood protection in New Orleans. This chapter is about a success story (at least, so far): the protection of London from flooding associated with tidal surges via the construction of the Thames Barrier. In particular, it highlights the role of one individual, Sir Hermann Bondi (intermittently, a senior scientific advisor to Her Majesty's Government from the mid-1960s to the early 1980s, and one of the United Kingdom's most famous 'mandarins' – the powers behind the politicians) in pushing through the project.

THE THAMES BARRIER

The Thames Barrier (Figure 8.1) was not started until 1974 after interim bank raising in 1971–72 (the Thames Barrier and Flood Protection Act was passed in 1972; this was needed to overcome institutional resistance) and not completed until October 1982 (and not officially opened until May 1984). The Thames Barrier has been described as the eighth wonder of the world. It is certainly a very impressive work of engineering. It consists of ten movable gates pivoted and supported between concrete piers (half a million tons of concrete were used during the construction of the Barrier), and powered electrically by hydraulic power packs. Six of the gates are rising (four of them being very large with a span of 61 meters each, two more about half that size), while the smallest four are falling gates with non-navigable openings close to the river bank. The main gates are hollow steel-plated structures, with the four largest more than 20 meters high (equivalent to a five-storey building), weighing about 3700 tons each, and with openings as wide as Tower Bridge. When not in use they rest on the

Source: Photo by Chang-Hee Christine Bae

Figure 8.1 The Thames Barrier

riverbed in semicircles of recessed concrete. In the event of a tidal surge requiring closure of the Barrier, the rising gates are moved up about 90 degrees and the smaller gates are brought down, resulting in a continuous steel wall. The Barrier spans 520 meters across the Thames at Woolwich Reach, very close to the Royal Docks.

Construction of the Barrier and associated works cost around £535 million (£1.3 billion at 2001 prices). Taxpayers (via the Ministry of Agriculture, Fisheries and Food) met 75 percent of the costs while ratepayers (that is, property taxpayers) were responsible for the remainder. The Barrier has been closed to protect London from the risk of flooding more than 80 times since 1982, with an upward trend in the incidence of closures (Figure 8.2).[1]

The Barrier is backed up by several flood defenses, primarily raising the banks (for example 80 kilometers between Putney and Purfleet in 1971–72, 32 kilometers downstream including the 60 meter-high Barking Barrier, and upstream of Putney on the South Bank and Hammersmith on the North Bank). In total, there are about 330 kilometers of riverbank defenses, many of them going back to the nineteenth century. It would have

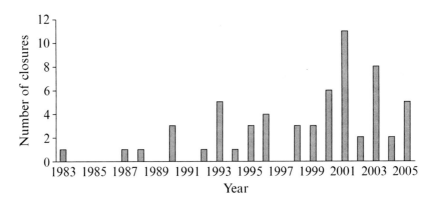

Source: Environment Agency.

Figure 8.2   Thames Barrier closures, 1983–2005

been possible to build a higher cost Barrier with much higher walls that would have been more permanent and reduce maintenance costs, but in addition to the cost disadvantage they would have destroyed the views and beauty of the river line.

MONITORING AND OPERATION

The costs of operating and maintaining the Barrier and the complementary riverbank defenses are about €11 million (at 2001 prices) with more than 80 employees. The monitoring system is almost foolproof, with the possible exception of a national (or even European) computer and telecommunications blackout (however, a well-researched recent disaster novel – Doyle, 2003 – outlines a scenario in which a 20 000-tonne tanker crashes into one of the Barrier gates, overwhelming London with 7000 cubic meters of water per second).[2] The combination of the Meteorological Office's Storm Tide Forecasting Service and the Barrier's own monitoring station tracks tide surges from the far north of Scotland down to the Thames (a process that typically takes 17 hours, although a 36-hour forecast is supplied) with information from satellites, oil rigs and land-based meteorological stations. The decision to close is made by the Barrier Controller, and usually takes place 3–4 hours before the peak surge (after which the Port of London Authority informs shipping and navigation lights are turned on).

THE GREAT FLOOD OF 1953

The background to the construction of this key engineering project starts from 1953. Although there have been reports about flooding in London stretching back 1000 years, the last time Central London experienced serious flooding was 1928. The 1953 'Great Flood' never reached Central London.

Tide levels are rising in the Thames Estuary by an estimated average of 60 cm per century (although the range of predictions is wide, between 26–86 cm by 2080). A tidal surge can be a threat under certain climatic conditions, particularly if a trough of low pressure, moving across the Atlantic, passes the north of Scotland then turns southward into the shallow waters of the southern North Sea, a problem that can be aggravated by strong northerly winds and coinciding with a high spring tide.

The recent history of coastal flood protection in the countries in the North Sea region can be traced to the 1953 flood. During the night of 31 January–1 February an unexpected storm-tide flooded the low-lying coastal areas of the countries around the North Sea. The resulting disaster in terms of loss of life and damage to infrastructure was significant. In the Netherlands, 1835 persons died; in the UK and Belgium, the loss of life was 307 (100 in East Anglia and 58 on Canvey Island) and 22 respectively. Most of the impacted area in the Netherlands was below sea level. In the UK 32 000 were evacuated, including five-sixths of the population of Canvey Island (10 000 out of 12 000). About 160 000 acres of Canvey Island were flooded. Most of the deaths occurred among the old, the very young and the disabled, especially in single-storey bungalows or prefabricated houses. Utilities (for example water, sewage, gas, electricity, telephone and rail services) were all disrupted.

Low-lying communities in the British coastal flood plain are protected against storm surge floods by sea defenses, such as permanent barriers, temporary gates and higher embankments. Coastal high–low tide differentials are typically around 4 meters, but high-water surges tend to be relatively short (1–3 hours). The extent of inland flooding is very sensitive to the volume of water passing through the defenses.

The flood zone (Figure 8.3) of the Thames Estuary contains one and a quarter million people and €80 billion of property spread over 45 square miles. The zone also includes 400 schools, 16 hospitals, 30 mainline rail stations, 68 subway and DLR (Docklands Light Rail) stations and 8 power stations, and the Port of London (with annual revenues of £2.7 billion). The business interruption cost of a major flood in the Thames floodplain (especially London) could be as high as $55 billion, without counting physical structure damage and the human costs in terms of lives and injuries.

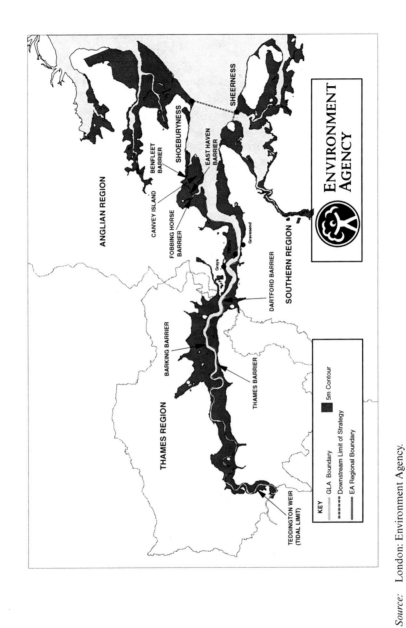

Source: London: Environment Agency.

Figure 8.3 Planning for Flood Risk Management in the Thames Estuary

The main obstacle to a barrier project at the time was that the volume of shipping using London Docks was at its historical peak, and that ships were getting bigger. This meant, according to the navigation experts, that an opening in the barrier of around 1400 feet (420 meters) would be required. A number of schemes were put forward, but failed to come to fruition. Then the whole system of sea transportation began to change. Cargo began to be shipped in containers on purpose-built ships, and a new container port was opened downstream at Tilbury. The old London Docks gradually became more or less redundant. It was then theoretically possible to restrict the width of the openings and locate the Barrier further upstream than had originally been intended; as we shall see, the final result was a little different.

After the 1953 floods the government appointed a committee to develop anti-flooding measures. Several alternatives were discussed and only in 1965 was government consent given to build a barrier. Sir Hermann's appointment (as a committee of one) to recommend a solution in October 1966 marked a turning point. The ultimate design by Charles Draper was chosen from 41 proposals.

THE ROLE OF SIR HERMANN BONDI

Sir Hermann Bondi (hereafter referred to as HB) was an academic and a world-class astronomer and mathematician who moved in and out of public life so many times and in so many different posts that he qualifies for 'mandarin' status. The standard image of the mandarin is of some bland, if wise, wizened old civil servant buried in some obscure office in Whitehall. HB did not fall into this category. A few quotations might give you a flavor of the man. Explaining the variety of his interests, he pointed out: 'A rut . . . is just like a grave, only longer' (Bondi, 1990, p. 128). As for success: 'Success is because of brilliance and education or sheer luck. I was lucky.' We don't think so, for someone who completed a two-hour college exam in 10 minutes. As for luck, he once commented: 'I am a very lucky man. I have been walking through life with a wide open mouth and roast ducks have come flying in with monotonous regularity' (Bondi, 1990, ix). Also, about luck: 'The worst luck is for a wicked fairy to wave her magic wand and make you the rich boy in a poor Jewish family.' When asked why he accepted the bureaucratic dead-end job as Director-General of the European Space Research Organization (ESRO) in 1967, he countered: 'It caters to my two vices – talking and traveling.' On the same lines, consider the last two sentences of his autobiography: 'Maybe all my career was a mistake. I should have become a travel agent' (Bondi, 1990, p. 137). A standard joke about his academic years was that he would accept any speaking invitation provided that the trip was more than 3000 miles. As for the range of his experience, the world's most extreme Anglophile was interned as an enemy alien in 1940 for 15 months[3] (shades of Japanese Americans), and he personally managed the septic tank for his house and designed its kitchen and heating system.

SOURCES

The next few sections of the chapter focus on the original decision and HB's role in moving it forward, especially in 1966. The major sources for this research are his personal papers stored at the Churchill Archives Centre in Churchill College, Cambridge, and his autobiography (Bondi, 1990). The point of the discussion is to emphasize some old-fashioned virtues long out of currency in planning circles: leadership, rational deliberation and top-down planning. No community participation, no stakeholder protection, no political quid pro quos, no US Congress shenanigans and no Louisiana-style corruption. Sometimes, perhaps not often, the old ways worked.

ANALYSIS

HB's discussion of risk–cost analysis would not be out of place in the relatively recent discussions in Congress and in the Department of Homeland Security (DHS) itself about how much to emphasize risk-based resource allocation. He would have looked at many of these discussions with great puzzlement, because anything but a risk-based approach made no sense. For him, the key elements were the damage associated with an event, the probability of that event occurring and the costs of prevention and/or dealing with its consequences. His main concern was that a tidal surge could penetrate the London subway system (deluging a large part of the system, closing it for up to several months, perhaps even a year). He was equally concerned about the potential loss of life because of not being able to evacuate the subway in time. He even proposed an unannounced trial evacuation despite the business interruption and the public complaints that this might have caused.

Also, he was not concerned about any complaints from other parts of the country that a diversion of available infrastructure resources to the Thames Barrier would deprive other regions of important projects. In his view, central London was the core of the national economy and any major interruption would be a 'knock-out blow to the nerve centre of the country' (Bondi, 1966, p. 7). Thus, London deserved and required a high priority analogous to the arguments about New York and Washington, DC in homeland security discussions in the United States.

Furthermore, although he did not have the tools that we have today to quantify the potential business interruption damages, he understood that they were so high that they might override the low probabilities of a disastrous event. The low probability of a cataclysmic event created many problems for him with outside institutions and other Barrier opponents. He had difficulty in persuading other involved public agencies to support the project. The Port of London used the low probabilities argument to try to avoid having to incur agency costs associated with the Barrier, not so much because of the project capital costs as for the annual increased maintenance costs (for example more dredging near their wharfs) that they would have had to pay. The antecedent agency of Transport for London was strangely unconcerned, although it had the most to lose. The risks were brushed away despite the tidal surge of 1953 (with its considerable damage and substantial deaths, not to mention the even more serious event near Rotterdam in the same surge), but much less severe than the event that the Barrier would protect against. The analogy to Katrina is quite clear. In other agencies, probabilities were based on the principle of using the past to predict the future, not basing it on the maximum damage event that might occur, primarily because the probability estimate of such an event was so uncertain.

HB's view was that while the timing of such an event was unpredictable, it was more or less certain to occur some day, and that protection was feasible. In that case, the key question for him was: how much would it cost to provide protection? He was not interested in exploring the realm of minutely probable events with minimal protection. Drawing upon his astronomer credentials (a co-founder of the steady-state theory of the universe that he later abandoned once the evidence for the big bang alternative became more compelling), he presented the following argument

> [Some disasters]: are completely unpredictable and completely unpreventable. If a very large meteorite were to strike the centre of London, the disaster would certainly be great, but nobody in his senses would ask the government to take action against such a contingency, because its probability of occurrence is unascertainably small and, secondly, because preventive measures do not exist. (Bondi, 1966, p. 4)

The tidal surge risk was not in this category. HB continues his case:

> With a foreseeable preventable disaster of immense magnitude, the situation is different. I think it is just as incumbent on the Government to prevent such a disaster as it is, for example, to prevent an enormous outbreak of smallpox. The precise probability of the disaster occurring becomes then a relatively unimportant matter.

If it can be shown, as I think it can, that the flood catastrophe to London would be of this immense kind, then, if we accept a hundred year period about which we wish to talk, I feel it is almost irrelevant whether the probability of such a disaster striking in the next hundred years is 10% or 1% or one-tenth of a per cent. This is the kind of thing that a community must not allow to happen if there are reasonable preventive measures and if the disaster is indeed of this enormous kind. (ibid.)

Is there any doubt that if HB had been in charge of the New Orleans levee decision at the time that it was made that he would have recommended protection against a category 5 hurricane event?

The Thames Barrier discussion focused on the degree of severity of the event that should be taken into account. The 1953 flood that tested the previous defenses was 3 ft 8 inches above the normal high tide level. A government report (the Waverly Committee) recommended a protection level of 6 ft or more higher.

HB recognized that technology was changing sea trade, which could have a major impact on the location of the Barrier:

Two such trends which are becoming very apparent indeed are the trend for an increasing proportion of goods to be carried by larger vessels, the rapid development of container traffic, and the trend that the intensive use of shipping will increase in the sense that the time of getting into port, remaining in port, and coming out of port, will assume an increasing importance. This implies that in all ports there is a tendency to provide new facilities for bigger vessels closer to the open sea. In the port of London itself, the construction of vast new facilities at Tilbury is a case in point and the gigantic construction of Europort outside Rotterdam is equally relevant . . . A further argument in this direction is the need for adequate transport facilities both by road and by rail to get goods to and from the port, and it is easier to provide these facilities not in a major existing city, but rather outside. (Bondi, 1966, pp. 11–12)

HB was particularly scathing about the primary report of the consulting engineers. They based all their analysis and scenarios on the assumption of a barrage 1400 ft wide, inconceivably without consideration of alternative locations. He was able to demonstrate that much narrower barriers (even as narrow as 350 ft) were feasible, but not necessarily desirable, at much lower cost (he also considered wider, downstream, barriers that could have been even more costly than the Barrier that was eventually built). There would have been many other implications from the siting choice in terms of maintenance costs, the risks of damage to ships and land value impacts. Also, many of the lower-cost options had to be ultimately rejected for political reasons; the government could not afford to risk the exposure of people to an event bigger than that of 1953. Taking into account siting considerations, the alternative barrier configuration and costs become very complex. Given the interdependencies among design features, location and cost, HB was able to cut through the complications by arguing that the key determinant was location, then to consider the pros and cons of alternative locations, and move on to evaluate design configurations and costs.

LOCATION

HB's critique of the consulting engineers' report is primarily based on the fact that they did not consider alternative sites for the Barrier. However, he focused on the key point: 'it seems to be reasonably clear that it would not be advantageous to have the site of the barrier higher than the Woolwich Reach of the river' (Bondi, 1966, pp. 24–5). Where was the Barrier built? At Woolwich Reach.

In general, the river widens as it flows downstream, so the Barrier would be wider, and presumably more costly, the closer its location to the estuary outlet. HB gave an exhaustive analysis of alternative sites, many of them downstream sites. We could discuss them, but it is unnecessary. His instinct was correct. In the final analysis, the cheapest, most narrow Barrier would have put too many people at risk. Also, a Barrier a couple of miles upstream would have jeopardized the historic borough of Greenwich, which has since been designated a World Heritage Site. However, cost considerations are not a trivial factor, so building the Barrier too far downstream would have been a mistake.

PROPOSALS FOR FUTURE PROTECTION

Now, a new, much more ambitious proposal is under consideration: building a 10-mile barrier from Sheerness on the Isle of Sheppey in Kent to Shoeburyness near Southend in Essex, in response to fears that the existing Barrier might not be able to cope beyond 2030 (see Figure 8.3). It would have a road on top linking the two counties, a step with significant travel time savings. Scientists for the Thames Estuary 2100 project, set up by the Environment Agency (the approximate UK equivalent to the US Environmental Protection Agency) to assess flood risk and river management over the next century, argue that the new barrier would be an option if sea levels rise as fast as expected. The existing Barrier will continue to be operational after 2030 but the standard of defense it offers will progressively decline.

Why the situation is deteriorating is a complex issue. The surge at the mouth of the Estuary (Sheerness–Southend) in 1953 was 2.59 meters. This was not the highest on record: the three highest were 3.66 meters in 1943, 3 meters in 1905 and 2.9 meters in 1894; they did not result in a disaster because they did not occur at a high spring tide. In 1953 the sea level rise was 4.69 meters. On the other hand, if you add the peak historical surge of 3.66 meters to the peak spring tide of 3.2 meters you get 6.86 meters, that approximates the peak capacity of the Barrier, 6.94 meters. The Barrier designers planned for a 5.5 meter rise (a 1000-year event); because of the amplification of the surge as the tide funnels up the river, they added 1.5 meters, 1.2 meters for the differential between London Bridge and Sheerness–Southend, plus additional wind and wave effects. If, instead of the 1.5 meter differential, you apply a plausible 35 percent upper river differential to the 5.5 meter Sheerness–Southend standard, you end up with 7.43 meters, that is, higher than the Barrier's protection.

The rise in the high-water level over time primarily reflects three factors: the increase in the global mean sea level; tectonic movement in Southern England resulting in sinking; and the effects of dredging, embanking and other human factors. Of these, the rise in the sea level is the most problematic. The Barrier designers assumed a 0.4-meter rise in the high-water level in the first 50 years, 0.22 meters of it because of the rise in the sea level. The currently accepted opinion is that a 0.31-meter increase in the latter is more likely because of global warming, regardless of the wide degree of disagreement about its possible causes. This results in the 1000-year event probability falling to 500 years in 2030 and 100 years by 2070. This is the thinking that governs the view that additional protection will be required by 2030.

As pointed out, in recent years the number of times the floodgates have been closed has been increasing. The Environment Agency estimates that the costs of additional protection will be about £4 billion up to 2025. If the current floodgates are breached, central London might be under 6 ft of water within a hour. A new flood risk management plan is being developed (Environment Agency, 2006), although the decision to build a new Barrier is unlikely very soon because of the availability of alternative near-term lower-cost options. It would be much more ambitious and more costly than the existing Barrier. The project would be the largest project of this type in the United Kingdom, although still smaller than some in the Netherlands.

It is important to note, however, that a new massive Barrier is an endgame option, the last resort rather than the first option. There are many steps that could be taken to provide additional protection after 2030. These include more building of higher banks and creating massive flood storage areas, initially using open space but later by abandoning some already developed lowlands and housing the displaced population elsewhere, for example at regeneration sites.

A major factor influencing the decision is a massive regeneration project, the Thames Gateway, which is planned to involve 200 000 residential units among other components, almost all of it built on the floodplain. It is a problematic idea. There is an intense housing shortage, especially in South-East England, but there are other potential and safer sites such as the release of some of the expansive greenbelt land or, even more dramatic, the more recent proposal of closing down Heathrow Airport to make room for more housing and to build a new international airport elsewhere. Where? Eye-poppingly, the suggestion is in the Thames Estuary, although there are potential airport sites on higher land rather than in the flood zone.

IMPLICATIONS FOR KATRINA

One of the most important elements of this narrative is its potential implications for a post-Katrina world in the United States. We would like to conclude by exploring a few:

1. Forgive the elitist position, but there are some situations in which pussy-footing around and tiptoeing through the tulips and all those similar, if mismatched, metaphors are the enemies of the good. Community participation, stakeholder involvement and political consensus are ideal in some circumstances, but not all. The concept of a Katrina czar was compelling. There are situations where expertise (or at least the ability to understand the experts) and leadership count. Without HB's drive and insight, the Thames Barrier would never have been built. Most of the stakeholders were opposed to the project, but he had the complete trust and support of the UK government at the time. The Environment Agency considers him its great hero.
2. We are certain that HB would have recommended building the levees in New Orleans (both then and now) to cope with a category 5 hurricane. He would have argued that the unique amenities of New Orleans justified it. This might not pass muster on strict cost–benefit grounds. However, HB would have dissected the problem in the following way: the costs of another Katrina would be huge, in societal as well as monetary terms; the probabilities are uncertain, but another Katrina is quite possible and perhaps likely, so the estimation of probabilities is almost irrelevant; and the consequences are preventable.[4]

    In addition, he would have rejected the notion that past experience can be used to predict the probabilities of future events (there are analogies here with earthquake research). Central London has not been impacted by a tidal surge since 1928 and in that event only 15 died; earlier events were reported in 1009, 1236 and 1663 although the nature of contemporary reporting obviously makes it difficult to assess the scale of the damage. However, HB's position would probably have been to assess the maximum damage event and evaluate whether it could be prevented at a tolerable cost. He certainly favored risk–cost analysis, but in this case the risks are unknown (not in terms of the incidence of tidal surges but in terms of their scale, especially in light of the fact that determinants of scale are dynamic and constantly changing).

3. However, he would have been very concerned about costs, and would have demanded an evaluation, monitoring and auditing system in place that would avoid the corruption for which Louisiana is famous.

4. His disparagement of the engineers' reports in the Thames Barrier example would have made him very skeptical about relying on the US Army Corps of Engineers alone. He would have demanded a second opinion, and probably a third.

5. We found little evidence in his papers that he had a specific interest in financial matters, but we believe that his view would have been that this is a national issue for which the costs should be shared, hence more of federal rather than local financing. His rationale, we suspect, is that the benefits of New Orleans' uniqueness to American society more than compensate for the costs of protection, given the homogenized America to which we are accustomed.

6. Are the implications of the Thames Barrier decision applicable to the pre-Katrina or the post-Katrina situation? The answer is both. The levees in New Orleans should have originally been built to category 5 hurricane standards, and if New Orleans is to be restored to a state approximating its original condition the reconstruction should be to category 5 standards. However, there is another important issue here, analogous to the Thames Gateway. Given the huge capital and maintenance costs, is it worthwhile to protect and restore such a vulnerable site?[5] As Friedrich Engels once said (although in a very different context): 'All great cities must perish.' Or, could the wonders of New Orleans be recreated (or, at least, metamorphosized) at another less beautiful inland site? We raise these questions, but we will refrain from answering them.

NOTES

1. The graph is from the Environment Agency. An answer to a question in Parliament on 16 June 2005 gave a somewhat higher number, 92 closures since the Barrier was opened

in 1982. The discrepancy is probably based on differences in the inclusion of particular reasons for closure. Regardless of the specific numbers, the smoothed trend is upwards.
2. The vulnerability of the Barrier is unclear. The stainless steel housing is not very strong, but it would take a large bomb to damage the gates, the concrete casings and the heavy machinery permanently. Moreover, for maximum impact, an attack would need to be carried out when the gates were closed at a time of unusually high tidal surge. It would require a high level of technical knowledge and long-term and continuous readiness to mount a successful attack of this kind. Of course, a symbolic attack with minimal damage is a different story. The level of security is modest: a steel gate, a security guard, and a check-in and screening station. However, the Barrier is closed to the public, although there is an adjacent public Information and Learning Centre.
3. For a parallel, Walter Isard, the grandfather of Regional Science, was a conscientious objector during World War II and was assigned as an orderly nurse in a mental hospital in Massachusetts. His response: 'I kept myself sane by translating August Losch.' In similar vein, HB gave impromptu lectures to his fellow internees on vector geometry.
4. His views were echoed in a recent media interview by Sarah Lavery, who is in charge of the Thames Estuary 2100 Project at the Environment Agency: 'There is a very small probability of the Thames flooding, but if it does the consequences are enormous. New Orleans is a nasty reminder to us all of the difference between the probability of something happening and the consequences if it does' http://news.bbc.co.uk/go/pr/fr/-/2/hi/science/nature/5092218.stm.
5. We do not doubt HB's answer. This gregarious, larger-than-life, Renaissance man would restore New Orleans almost, we suspect, at any cost.

REFERENCES

Bondi, Hermann (1966), *London Flood Barrier*, London: Ministry of Housing and Local Government.
Bondi, Hermann (1990), *Science, Churchill and Me*, Oxford: Pergamon Press.
Doyle, Richard (2003), *Flood*, Toronto: Random House.
Environment Agency (2006), *Planning for the Next 100 Years of Flood Risk Management for London and the Thames Estuary*, Report, London: Her Majesty's Treasury.

9. Is New Orleans ready to celebrate after Katrina? Evidence from Mardi Gras and the tourism industry

Kathleen Deloughery

INTRODUCTION

Once the full effect of Hurricane Katrina was realized, many people recognized that the recovery of the impacted area, both physical and economic, would be a long process. With almost 70 percent of the population still evacuated from the city a month after Katrina and with many of those already building new lives in another area, some wondered if New Orleans would ever return to what it recently was. However, since the Hurricane, officials and citizens have worked hard to rebuild New Orleans, and even attract visitors to the area once again. One aspect of New Orleans that might have made this rebirth easier was that the city had a specific goal in mind: to host Mardi Gras in 2006. Mardi Gras is an iconic event and hosting it was seen as important to the Louisiana economy; it is a time when many tourists come to New Orleans and surrounding areas. In 2006, the celebration would be even more important, because the national spotlight would once again fall on New Orleans. Outsiders might see a signal that New Orleans had made significant progress in overcoming the devastation brought by Katrina.

Before Hurricane Katrina, tourism was Louisiana's second-largest industry and private employer, directly accounting for roughly 4 percent of the state's gross state product.[1] The importance of resurrecting the tourism industry in Louisiana can be deduced from the number of commercials the state has created trying to encourage people to 'rediscover' New Orleans. The prosperity of the tourist industry plays a vital role in the full recovery of New Orleans. As Mardi Gras fell in late February in 2006, with festivities typically starting at least ten days beforehand, that left the city just over five months to become operational and attractive to tourists again. This chapter will examine two questions. First, did the looming Mardi Gras celebration help hasten the pace of the recovery effort in New Orleans?

Second, did Mardi Gras provide a large and powerful enough shock to the New Orleans economy to help aid the economic recovery of New Orleans?

DATA

This discussion will look at several aspects of economic recovery. First, key indicators of recovery for both the Orleans Parish and the New Orleans Metropolitan Service Area (MSA) will be evaluated. Orleans Parish covers the city of New Orleans, while the New Orleans MSA covers Orleans Parish and six other surrounding parishes.

From the indicators in Table 9.1, we see that the recovery of New Orleans is lagging somewhat behind the recovery of the entire MSA in all major categories including population, open public schools and open hospitals. That is probably because Orleans Parish was the hardest hit of the seven parishes in the MSA.[2] For instance, the MSA reached almost 70 percent of its pre-Katrina population within a year. On the other hand, the city of New Orleans has only seen 40 percent of the people previously living in New Orleans return, with city officials estimating that no more than 60 percent will return in the long run. In fact, some of the parishes in the New Orleans MSA have experienced modest population growth since Hurricane Katrina. It could be that some individuals unable to move back to Orleans or Jefferson Parish have settled in a neighboring parish that was not damaged as extensively. Additionally, while over half of the hospitals and public schools in the MSA had reopened by March, that number was less than 30 percent for the city of New Orleans. Despite more than half of the hospitals in the MSA reopening, the bed capacity in the area is still only at 45 percent of its pre-Katrina level. These indicators show that, while the area has recovered somewhat since September 2005, there is still far to go.

In the rest of this chapter, I will examine available data that yield information on the tourism industry in New Orleans. These data consist largely of hotel occupancy and airport passenger data from 2000. These data will be used to predict how long it will take the tourism industry in New Orleans to recover to its pre-Katrina level of activity. Usually, data on average expenditure per visitor would be important in this analysis. Additionally, a key New Orleans goal was to show the world substantial recovery by the Mardi Gras celebration, so as to create a powerful signal. Therefore, I will compare the size and scope of the 2006 Mardi Gras celebration to those in the past. Finally, I will examine the impact Mardi Gras had on the tourism sector in 2006 by comparing the percentage change in activity from the same time period the year before to see if Mardi Gras boosted activity in

Table 9.1   Basic data before and after Katrina

	July 2005	Sept 2005	Mar 2006	% of pre-Katrina activity	% change growth
Population	1 292 774	703 387	887 879	68.68	14.27
Population**	437 186	138 681	181 400	41.49	9.77
Employment (non-farm)	610 200	405 500	425 700	69.76	3.31
Employment (non-farm)**	406 583	114 134	138 238	34.00	5.93
Construction employment	30 100	12 700	19 100	63.46	21.26
New housing permits[1]	417	280	486	116.55	49.40
Open public schools	317	31	189	59.62	49.84
Open public schools**	122	0	21	17.21	17.21
Hotel employment	15 800	9 000	11 500	72.78	15.82
Restaurant employment	56 300	28 200	37 900	67.32	17.23
Sales tax collections[2]	64 712 092	30 788 490	91 438 713	141.30	93.72
Airport traffic	881 552	42 198	483 176	54.81	50.02
Open hospitals	53	17	31	58.49	26.42
Open hospitals**[3]	25	0	7	28.00	28.00

Notes: ** Data for Orleans Parish only
1. Housing permits fell again after rebuilding codes were announced (April)
2. Sales tax collections usually rise after a disaster
3. Only 45% of hospital beds are open

Source: Data from BLS, Louis Armstrong International Airport, Smith Travel Research, and http://www.lorenscottassociates.com.

the hotel or airport industry. Additionally, if there was a boost in activity, I will look at whether or not that impact is expected to persist.

AIRPORT RESULTS

First, data from the Louis Armstrong International Airport in New Orleans will be examined. These data include all passenger enplanements and deplanements from January 2001 until May 2006.[3] Figure 9.1 includes passengers on both domestic and international flights.

There is some seasonal variation in travel activity, as can be seen in the figure. Much of the variation can be explained as monthly fluctuations in

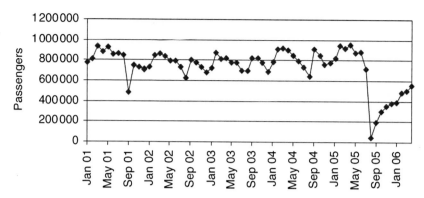

Figure 9.1   Passenger activity at Louis Armstrong International Airport in New Orleans January 2001–present

the number of visitors. The summer months are notoriously the slow season for tourists, while October and February through April usually see the most tourist activity. Once we control for monthly fluctuations, most additional movements can be accounted for by the number of visitors in the previous month. In other words, we can model tourism as an auto-regressive process with seasonal variation. The big drop-off in passenger activity occurred in September 2005, just a few days after Hurricane Katrina hit. Since then we have seen a steady increase in the number of passengers traveling through New Orleans, but that number has yet to reach the level of any pre-Katrina month with the exception of September 2001, which is an outlier itself. The number of passenger-carrying flights has followed a similar trend.

With that in mind, it could be useful to predict the length of time it would take to return to 100 percent of pre-Katrina activity. In order to do that, we calculate the current deviation from pre-Katrina means. This calculation can be done in the following manner. First, we take the number of passengers traveling in month *n* since September 2005. Then we divide that by the average number of passengers traveling in that same month before September 2005. These calculations yield the trend shown in Figure 9.2.

According to the top graph in Figure 9.2, the recovery of New Orleans Airport is following a logarithmic trend[4] and returned to 63 percent of its pre-Katrina activity. This same calculation was done for the number of passenger flights. Initially, the number of passengers increased faster than the number of passenger flights. However, later passenger flights recovery pulled even with that of number of passengers. In May 2006, the number of passenger flights had also returned to 63 percent of pre-Katrina activity. The level of recovery in the two indicators was not separated by more than

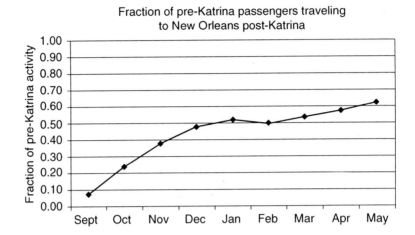

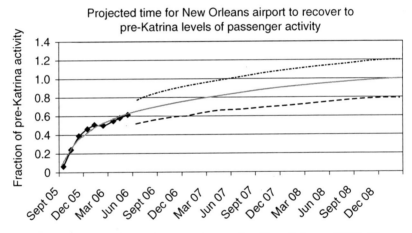

Figure 9.2   Pre- and post-Katrina air travel to New Orleans, 2005–06

10 percent in any month. This relationship is to be expected since airlines want to choose the number of passenger flights based on the projected number of people who will be flying. Southwest Airlines, which has the most passenger flights out of New Orleans, has said it does not expect to add any additional flights until almost all the 'cruise ships and conventions return to New Orleans'. Therefore, we will focus solely on passenger activity. In order to determine how long it will take for the airport to return to 100 percent of its pre-Katrina passenger activity, we fit these data to a logarithmic regression as shown in the bottom graph of Figure 9.2.

According to this trend, we can expect the airport to return to its pre-Katrina activity by January 2009. However, because of the small number of data points, it is important to look at the confidence interval as well as just point estimates. Therefore, Figure 9.2 also includes 95 percent confidence intervals that have been allowed to vary with time so that our point estimates become less precise as we get farther away from the present. According to our confidence intervals, the full recovery of the airport could be as late as 2016. Keep in mind that these estimates all assume that no more natural disasters befall New Orleans.

HOTEL RESULTS

Next, I will focus on data on area hotel occupancy.[5] Many hotels in the area faced the same problems as individual residences: property damage from strong winds and flooding. The number of available hotel rooms, after showing a slight upward trend of about 1300 new hotel rooms each year over five years to a total of over 38 000 rooms before Katrina hit, fell drastically in September 2005. After falling to a low of just over 11 000 rooms, the industry rebounded in October to almost 25 000 rooms. Hotel room closings have come from two main sources. First, some hotels closed their doors entirely, at least temporarily, after Hurricane Katrina. For instance, only 68 percent of all the hotels that were in New Orleans in August 2005 had reopened a year later. Second, some hotels were not able to make all their rooms immediately available, either due to flood damage on the lower floors or wind damage on the higher floors.

Interestingly, hotel occupancy rates have stayed fairly stable between 2000 and 2006 including the time around Hurricane Katrina. However, those estimates are misleading because after Hurricane Katrina, the New Orleans tourism industry experienced a shift in both supply and demand. Since fewer people were coming to New Orleans and there were fewer hotel rooms available, occupancy rates could be expected to stay fairly stable. However, once the number of available rooms is corrected for, occupancy rates can be used to calculate the number of occupied hotel rooms (Figure 9.3).

Again, we can see a large drop off in the number of occupied rooms in September 2005. Since then, we have seen more recovery than in airline passenger travel. However, since Katrina hit New Orleans, no single month has reached the average number of pre-Katrina hotel occupancy, which was just over 23 000 rooms. To estimate how much time is necessary for the hotel industry to recover fully, I implement the same framework as I did with hotel data to generate Figure 9.4.

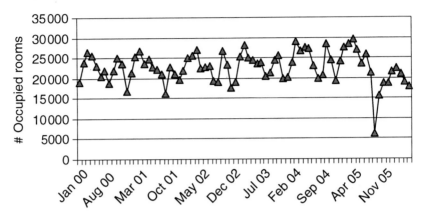

Figure 9.3   Number of occupied rooms in New Orleans

The upper line shows the number of occupied hotel rooms in month *n* after Katrina hit divided by the average number of occupied hotel rooms in month *n* before Katrina hit New Orleans. Unlike the airport data, these data do not yield a simple logarithmic trend. Therefore, we must explore reasons to explain why in December and January hotels were doing more business than in the previous year. Additionally: why have we seen a strict downward turn recently? Luckily, a few testable explanations are readily available.

First, the population of New Orleans has been on the rise since October 2005. Many of the people who were returning to New Orleans did not immediately have a home to go back to. Therefore, some of these people might have stayed in hotels when they first moved back, then slowly moved out of the hotels into their homes. Therefore, we should control for the number of people moving back into New Orleans in each month. First, I controlled for total population. In Figure 9.5, the upper line plots the number of occupied hotel rooms after controlling for the current population of New Orleans. One can see that this does not help explain away any salient features of the shape of the curve. Next, I control for the number of people moving back in one month instead of the current population because people returning to New Orleans are probably not permanently staying in hotels. Instead, they may stay in the hotels for anywhere from a week to a month while their home is being rebuilt. This is harder to look at graphically, so instead we must perform the calculation numerically. Therefore, I ran a regression with deviations from the trend as the dependent variable and used changes in population from the previous month as the independent variable. This regression yielded an R^2 of 0.026, which is very low. This finding implies that change in population has almost no

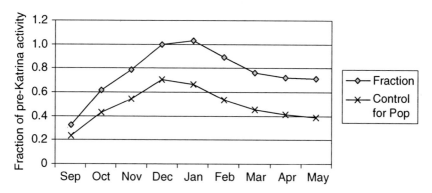

Figure 9.4   Deviations from trend in occupied hotel rooms in New Orleans

explanatory power in accounting for the deviations from past trends in occupied hotel rooms in New Orleans. The second testable explanation is that many of the people staying in New Orleans are relief or temporary workers. While some of these individuals are signing year-long leases with apartments,[6] those staying for a shorter period of time might be living in the hotels. Unfortunately, a monthly estimate of the number of relief workers in New Orleans is not available. If this information were available, controlling for that population would enable us to get at the true deviation from trend in visitors to New Orleans staying in hotels and predict how long it would take the hotel industry to recover.

MARDI GRAS RESULTS

In order for tourists to return to New Orleans, the city's residents must be active in the recovery effort. Mardi Gras is one of the events that draws tourists into Louisiana, and New Orleans especially. Additionally, Mardi Gras usually receives news coverage on major outlets, suggesting that the celebration will be viewed by others deciding whether or not to visit New Orleans in the future. Therefore, a successful Mardi Gras celebration is invaluable to the recovery of tourism in New Orleans. Thus, Mardi Gras should be a priority for the current residents. The main avenue for locals to participate in Mardi Gras is through the parades. Adult social organizations, called Krewes, are in charge of putting on the parades during Mardi Gras. Their responsibilities include the timing and location of the parade, the number of floats and throws,[7] and securing bands and other entertainment. Therefore, looking at the history of participation of the Krewes in

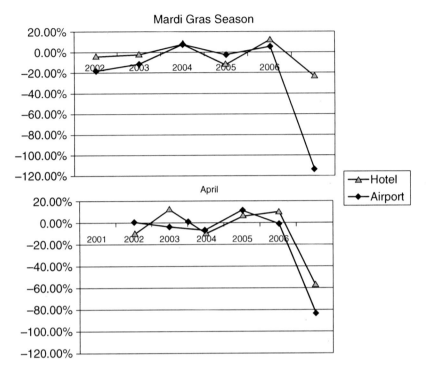

Figure 9.5   Percent change in airport and hotel activity

Mardi Gras will lend itself to inferences about local participation in Mardi Gras after Hurricane Katrina. Before Katrina hit New Orleans, the number of Krewes participating in Mardi Gras stayed constant at 33. Here, participating in Mardi Gras means that the Krewes hosted their own parade that went through Orleans Parish. Additionally, the average number of floats in each parade stayed relatively stable at just over 22. However, after Hurricane Katrina hit, many Krewes suffered a loss in membership because people moved away. Therefore, they did not have the resources to put on the same celebration as in past years. In fact, four Krewes decided not to participate in Mardi Gras at all in 2006. An additional three Krewes decided to join together to host one parade instead of the three separate parades they had put on in the past. In addition to there being six fewer parades in Orleans Parish in 2006, each parade had, on average, three fewer floats. Therefore, local participation in Mardi Gras was lower than in years past. However, the population of New Orleans, and the membership count of the Krewes, is much lower than it was pre-Katrina. Nine of the 26 Krewes that hosted their own Mardi Gras parades in 2006 released information on the number of members they had in 2005 versus 2006 during the Mardi Gras celebration. For those Krewes, membership fell by about 27 percent after Hurricane Katrina. Similarly, the number of floats these nine Krewes provided for their respective parades decreased by 26 percent. If we can consider these Krewes to be representative of the whole group, then the scaled-back number of floats in each parade can be thought of as a direct result of the number of people moving out of New Orleans.

Of course, locals can participate in the planning and execution of Mardi Gras, but the celebration will still not be a success if tourists do not visit. The tourism bureau estimated that 1 million people usually attend the last weekend of the Mardi Gras Celebration. In 2006, only 700 000 attended, or about 70 percent of the pre-Katrina level. The good news is that Mardi Gras does not appear to be a huge outlier in terms of large tourist attractions in New Orleans. The city usually hosts two other festivals each year, the Jazz and Heritage Festival and the Essence Festival. The Jazz Festival usually draws a crowd of just over 500 000, but in 2006 only 350 000 spectators attended the festival in late April to early May. Again, this is just under 70 percent of the pre-Katrina level of participation. On the other hand, the Essence Festival decided to move to Houston, so New Orleans did lose an opportunity to attract more visitors with that move.

Finally, many in New Orleans thought that Mardi Gras would be the mechanism that would reinvigorate the tourism industry, and thus the economic recovery, of the area. In order to determine what type of catalyst Mardi Gras was able to provide, we will look at the percentage change in activity in the number of occupied hotel rooms and airline passengers during the Mardi Gras Season between 2000 and 2006. Because the timing of Mardi Gras is tied to the Easter holiday, it occurs on different days each year, ranging from early February until mid-March.

Prior to Hurricane Katrina, one can see that there were slight fluctuations each year in hotel and airport activity. Additionally, these fluctuations usually moved in the same direction, lending credence to the belief that they are correlated. However, after Hurricane Katrina hit, we see that the airline industry experienced a much sharper drop-off from the previous year than the hotel industry did. There are several possible explanations for this phenomenon. First, it may be that the hotel industry did not experience as much damage as the airline industry. Additionally, it could be that in February the number of occupied rooms is still not perfectly reflective of the number of tourists in the region; instead we are picking up relief workers and those waiting to move back into their homes. Finally, it could be that Mardi Gras provided a boost to the hotel industry, but was not able to provide that same boost to the airline industry. Individuals who went to New Orleans for

Mardi Gras would obviously need a place to stay regardless, thus the hotel industry benefited from the visitors. However, it is possible that more groups decided to drive instead of fly because they might not have been knowledgeable about the current condition of the airport or possibly due to the high cost of airline tickets.

In order to examine these different explanations, I looked at the same calculations and graphs for a random month. Information on May, June, July and August is not available for both industries. Additionally, data from September are biased by both the evacuations of New Orleans and by the events of 11 September 2001. Therefore, I repeated these calculations for April. I looked at the percentage change in activity in both the hotel and airline industry in April of each year between 2000 and 2006. Note that there are several key differences in this graph and the previous one. First, even before Hurricane Katrina, the airline and hotel industry did not always move in the same direction in April (that is, one would experience growth over one year while the other would experience a decline). Additionally, we do not see the huge divergence between the hotel and airline industry that was evident after Katrina in the Mardi Gras case. However, we must proceed carefully in trying to draw any conclusions from these findings. For instance, it is premature to say that Mardi Gras provided a short-lived boost in helping the hotel industry recover. There are several reasons that these conclusions are dangerous. First, we only have five years of data, and only one year after Hurricane Katrina hit. That is not enough information to draw strong conclusions. However, as time passes, we can continue asking these questions and hopefully get better estimates, but right now we must work with the data we currently have. Second, it is possible that during the Mardi Gras season there were still more temporary workers living in hotels than there were in April, which could skew the results. If this were the case, then we would expect to see a large divergence between airline activity and hotel activity early on in the recovery effort and for this gap to lessen over time.

CONCLUSIONS

Considering the level of destruction and devastation seen in the first year after Katrina in Louisiana, and especially New Orleans, the recovery efforts made have been effective. State and city officials realized that in order for the area to continue to thrive, tourists once again have to come back to New Orleans. Advertising campaigns have encouraged visitors to come and 'rediscover' New Orleans while the city has tried to go about 'business as usual'. Their efforts seem to be working. By the summer of 2007 the airport was functioning at 72 percent of its pre-Katrina activity, and has experienced growth back toward the mean in every month. The hotel industry is expecting a large return of tourists, as evidenced by the decision to add another 1150 hotel rooms to the downtown area. By the summer of 2007 the number of occupied hotel rooms was over 70 percent of the pre-Katrina average. However, these numbers may be biased by the number of temporary workers staying in hotels while they work in New Orleans. Therefore, more information must be processed in order to tease out the true number of hotel rooms occupied by tourists.

In fact, New Orleans was able to recover so quickly that within five months it was able to host successfully a Mardi Gras celebration that drew in 70 percent of the normal crowd size. Both the crowd and the celebration were scaled down, but that is to be expected. Actually, participation in Krewes, the number of floats per parade, and the amount of visitors to the celebration all fell by about 30 percent. Additionally, according to our crude data, it appears that Mardi Gras has possibly provided a positive shock to the hotel industry in comparison to the airport industry. Again, that finding is still premature until we can get a better estimate of the number of tourists in hotels.

Finally, all estimates in this chapter will become more precise as more time goes by. Hurricane Katrina hit so recently that we do not have enough data points to extrapolate information from since September 2005. Especially when we try to look at information from Mardi Gras, more years of information will make our analysis more precise. This chapter focuses on a detailed analysis of the 2006 Mardi Gras. The Mardi Gras of 2007 was even more successful with close to 800 000 visitors and sales tax revenues returned to 84 per cent of pre-Katrina levels (City of New Orleans, 2007). Additionally, all estimates in this chapter assume that no more natural disasters hit New Orleans. Even though summer has always been a slow season for tourism in New Orleans, fear of hurricanes or disruptions from mandatory evacuations might cause slower recovery than we are predicting. Therefore, information for this chapter should continue to be gathered over the next several years to see if the trends realized within the first year continue.

NOTES

1. The Department of Tourism predicted that the tourism industry indirectly led to an additional \$4.1 billion of activity in Louisiana in 2004.
2. The MSA includes Orleans, Jefferson, Plaquemines, St Bernard, St Charles, St John the Baptist and St Tammany.
3. Data were received from Louis Armstrong International Airport.
4. The R^2 is 0.97, but that is probably higher than usual since we have so few data points.

5. Hotel data were bought from Smith Travel Research.
6. According to Reis's New Orleans Update apartments are at 100 percent occupancy, but landlords are estimating half of those are short-term workers, not returning tenants (Reis, Inc., 2006).
7. Throws consist of beads, dolls, medallions, and so on that are thrown from the float to the parade watchers.

REFERENCES

Blake, Nathan and Thomas Fomby (1994), 'Large shocks, small shocks, and economic fluctuations: outliers in macroeconomic time series', *Journal of Applied Econometrics*, **9**, 181–200.

City of New Orleans (2007), *Press Release, Office of Communication*, 16 August.

Duke, Rex, *Best of New Orleans – Gambit Weekly Newspaper* (Feb–Mar 2003–2006).

Horwich, George (2000), 'Economic lessons of the Kobe earthquake', *Economic Development and Cultural Change*, **48** (3), 521–42.

http:// www.bls.gov.

http://www.demographia.com/db-katrina.htm.

Lemons, Hoyt (1957), 'Physical characteristics of disasters: historical and statistical review', *Annals of the American Academy of Political and Social Science*, **309**, Disasters and Disaster Relief (Jan) p. 1–14.

Louisiana Tourism Satellite Account (2004), Louisiana Department of Culture, Recreation, and Tourism.

Reis, Inc. (2006), 'Reis's New Orleans update', First Quarter.

Sayre, Alan (2006), 'New Orleans desperately seeking tourists', Associated Press, 31 July.

Scott, Loren (2006), 'Advancing in the aftermath II: Tracking the recovery from Katrina and Rita', June.

Tsay, Ruey (1988), 'Outliers level shifts, and variance changes in time series', *Journal of Forecasting*, **7**, 1–20.

10. Estimating the state-by-state economic impacts of Hurricane Katrina

Jiyoung Park, Peter Gordon, James E. Moore II, Harry W. Richardson, Soojung Kim and Yunkyung Kim*

INTRODUCTION AND OBJECTIVES OF THE CHAPTER

Extreme natural events that strike major population centers are relatively few. When they do occur, the results are tragic in terms of human suffering and economic loss. Yet, if we can extract lessons that mitigate future losses from similar events, we will have at least garnered some benefits. We hope to accomplish some of this in the research reported here, by estimating and reporting some of the state-by-state economic losses associated with Hurricane Katrina.

The 1995 earthquake in Kobe, Japan, required massive infusions of funds but the area's economy rebounded relatively quickly (Horwich, 2000). It appears that large losses of physical capital are mitigated if human capital losses are small and if human capital is relatively mobile, both sectorally and spatially.

The second reason for our study is to test the expanded capabilities of NIEMO, the National Interstate Economic Model. NIEMO is an operational 50-state (plus the District of Columbia) multiregional input–output model that we have used to estimate the interstate effects of hypothetical terrorist attacks on the top three US seaports (Park et al., 2007). The application described in the following pages is to the loss of port services at the port of New Orleans and nearby harbors in the Louisiana Customs District in the first seven months following Hurricane Katrina. Work to estimate the additional impacts from tourism declines and gas and oil losses is currently under way. In the research reported here we have added a supply-side capability to the model. Export losses prompt demand-side multipliers while import losses prompt supply-side multipliers. We report the state-by-state losses of each.

NIEMO AND PREVIOUS APPLICATIONS

NIEMO Background

The detailed steps involve in constructing NIEMO are described in Chapter 11 of Park et al. (2007), so only a brief introduction is offered here. Following the suggestion of an ideal Interregional Input–Output (IRIO) model by Isard (1951), a Chenery–Moses-type multiregional input–output MRIO model was developed as an alternate and potentially operational approach requiring a more simplified data set than an IRIO (Chenery, 1953; Moses, 1955). NIEMO revives an approach adopted in the late 1970s and early 1980s (Polenske, 1980; Jack Faucett Associates, 1983), the development of an MRIO model. NIEMO is an operational state-level model of the US; it combines state (plus Washington DC data) from 51 IMPLAN input–output models with interregional trade flows based on the Commodity Flow Survey (CFS). Data from both are aggregated to 47 economic sectors (29 commodity sectors and 18 service sectors) that can be easily converted to other US industry codes. The model simulates production and trade between 52 regions (50 states, Washington, DC and the 'Rest of the World'). This results in an MRIO matrix with almost 6 million cells ($= (51 \times 47)^2$).

Building NIEMO involved substantial data assembly and considerable data manipulation. We used the state-by-state trade flows only for commodities, omitting trade in service sectors, which are not reported in the CFS.

There are at least two problems applying the CFS data; there are many unreported values as well as no information for the transactions of service sectors. Nevertheless, there have been three attempts to estimate interregional trade flows using the data from the 1997 Commodity Flow Survey (CFS). Jackson et al. (2006) used IMPLAN data to adjust incomplete CFS information primarily by adopting a Box-Cox transformation as well as double-log distance-decay functions. Another approach is to use a doubly constrained gravity model based on county-level data from IMPLAN and ton-mile data from the CFS (Lindall et al., 2005). The procedure suggested by Park et al. (2004) uses the same basic data sources, but adopts a different estimation approach to update the missing CFS data and a doubly constrained Fratar model (DFM) to update the 1997 trade flows to estimates for 2001.

After estimating the trade flows, NIEMO requires another set of basic data: industrial trade coefficients matrices for each state and Washington, DC. While the interstate trade tables by industry are difficult to construct because of incomplete information in the CFS, inter-industry tables present fewer problems because reliable data are available from IMPLAN at the state and industry levels. For details of the procedure used to estimate values for the empty cells in the trade flow matrix, see Park et al. (2004). A slightly less serious problem is that the currently available 1997 CFS data had to be updated to match the 2001 IMPLAN data. However, it is still problematic to rely wholly on the 2002 CFS data for a direct match with 2002 IMPLAN data because the sample size of the 2002 CFS was half that of the 1997 CFS.

Among NIEMO's many strengths, the greatest is the ability to estimate spatially detailed indirect economic impacts. Direct economic impacts are often estimated in the aftermath of an event. If, for example, plausible scenarios for the time-profile of reduced shipping facilities are available, spatially detailed indirect and induced economic effects can be estimated with a NIEMO-type model. Standard applications of input–output (IO) that determine indirect and induced impacts typically do not include interactions among industries and states. In this sense, the trial version of Park et al. (2006a), the demand-driven MRIO-type model that is described in the next subsection, provides a unique approach to an initial examination of the data.

While preliminary results to test NIEMO's accuracy showed that almost 6 million multipliers within NIEMO can be estimated at low cost, we have also shown that NIEMO's estimates are plausible (Park and Gordon, 2005). Further elaborations of the initial NIEMO model, estimating service sector interactions, are under way, using geographical weighted regression (GWR) methods (Park, 2006b).

Applying the initial demand-side NIEMO, Richardson et al. (2007b) examined the national and interstate economic impacts of terrorist attacks on major US theme parks, and Park et al. (2006) reported the economic impacts from foreign export closures resulting from BSE (bovine spongiform encephalopathy) in the state of Washington. Furthermore, the initial development and application of the supply-driven model can be found in Park (2006a).

Three-Ports Applications of NIEMO

The terrorist attacks on 11 September 2001 prompted several studies by economists to evaluate the socio-economic impacts on the US economy from hypothetical terrorist attacks. A recent study by Park et al. (2007) addressed the economic impacts of terrorist attacks on the major ports for domestic and foreign export disturbances. In this application, the authors utilized the initial version of NIEMO to estimate industry-level impacts from the short-term loss of the services of three major US seaports – Los

Angeles/Long Beach, New York/Newark and Houston – on the economies of all 50 states and Washington, DC, as a consequence of hypothetical radiological dirty bomb attacks. They treated the seaports of Los Angeles and Long Beach as one complex, LA/LB, the adjacent seaports in New York and Newark as NY/NJ, and analyzed the attacks on the three port complexes as alternatives rather than as simultaneous events. This is because 'it may be easier to plant simultaneous radiological bombs . . . on outbound rather than inbound freight, especially because the effects may not be very different if the bombs are set off at the perimeter (prior to passing through security) rather in the heart of the port terminal' (Gordon et al., 2005, p. 264).

Based on hypothetical export final demand losses for one month, USC-sector state-by-state indirect impacts from attacks on the three ports were estimated. The sum of indirect impacts of an attack on New York and New Jersey reached $2.75 billion, $2.64 billion for California and $2.23 billion for Texas, based on direct impacts of $4.69 billion for New York and New Jersey (NY/NJ seaports), $4.11 billion for California (LA/LB seaports) and $2.23 billion for Texas (Houston seaport). Total impacts in other states varied and were generally a function of state size and distance from the impacted port. When including the direct effects of import losses for the states where the attack takes place, multipliers summed across all states ranged from 1.24 (California) to 1.98 (Texas). The differences are accounted for by the fact that LA/LB has the largest value of imports; the authors suggested for further research the estimation of indirect impacts from the losses of imports by Park (2006a). The loss of the services of the LA/LB port would cost the US economy approximately $22.8 billion. Corresponding impacts for losses of port services in NY–NJ and Houston are $16.2 billion and $9.7 billion, respectively. However, it should be noted that if ports are unusable for longer periods, these losses would grow, although strict pro-portionality would be an overstatement of the impacts because substitution options become more feasible and important as time passes.

NIEMO simulations provide estimates of state-by-state impacts for all USC sectors. Specific results of sectoral effects for the five largest sectors in terms of total US output are also discussed in Park et al. (2007).

ELABORATION OF NIEMO TO SUPPLY-SIDE APPLICATIONS

In the classic IO system, there are two standard models. One is the 'Leontief' demand-driven IO model, following Leontief's initial efforts (1936, 1941) to generalize interdependences between industries in an economy. To address 'the highly complex network of interrelationships which transmits the impulses of any local primary change into the remotest corners of the economic system', the general static equilibrium for an economy was constructed (Leontief, 1976, p. 34). In the classic IO system, therefore, the interrelations between industries take account of all of the 'technical' relationships in an economy via constant coefficient production functions.

Two key assumptions implicit in the Leontief model – a competitive market system and non-scarce resources – were singled out by Ghosh (1958), who suggested another approach to understanding the interrelations between industries. The technical coefficients from the Leontief model are assumed to be fixed and yield new industry total outputs that change with final demands. This is thought to work 'so long as there is no scarce factor and so long as suppliers are able to offer more of any commodity at the existing price' (Ghosh, 1958, p. 59) in the short run.

Some criticisms of the supply-driven model were suggested by Oosterhaven (1988). He questioned whether based on given final demands, 'local consumption or investment reacts perfectly to any change in supply'; for example, if 'purchases are made, e.g., of cars without gas and factories without machines'. Using Taylor's expansion, he concluded that 'both as a general description of the working of any economy and as a way to estimate the effects of loosening or tightening the supply of one scarce resource, the supply-driven model should not be used', and suggested an alternative model instead of using the supply-driven model directly.

Two other studies by Gruver (1989) and Rose and Allison (1989) elaborated the critique. However, neither touched on the core debate over the implausibility of the model (Oosterhaven, 1989, 1996). Dietzenbacher (1997)'s interesting interpretation showed that the supply-driven IO model is equivalent to the Leontief price IO model, and that if the supply model is interpreted as a price model instead of a quantity model, the problems of implausibility vanish. This interpretation highlights the condition that exogenous changes in value added should be followed by price changes of value added, not just quantity changes. Hence, focusing on price changes similar to the Leontief price model, Dietzenbacher called the supply-driven model the 'Ghoshian' price model. Although this interpretation provides theoretical support for the supply-driven model, it is difficult to find empirical applications of supply shocks in the literature. This might be due to difficulties of interpretation that direct and indirect impacts by a disruption in the supply side are presumably quantity losses rather than price decreases. Current research shows that there is no adequate way to interpret the losses in the Ghoshian price model and focus on the forward linkages (Dietzenbacher, 2002) or structural changes of an economy (Wang, 1997; Bon and Yashiro, 1995; Bon, 2001) using the supply-driven model.

However, as Ghosh (1958) suggested, we assume that producers will not decrease their previous outputs or factors during short-term periods even if there are significant shocks to the economy. Finding substitute products involves various costs, for example costs for searching for substitutes, additional transportation costs, and so on, and hence unless those are expected for the long run, their reactions will be delayed.

Even in the case that a normal equilibrium is not maintained, Ghosh's quantity model can be applied as a Ghoshian price model, and play a role in estimating economic impacts. To implement this switch, it is necessary to introduce exogenous price elasticities of demand and combine them with the supply-driven model, adjusting quantity responses to price impacts. Using an iterative adjustment based on Taylor's expansion of the supply-driven model yields better estimates than simple linear multiplicands. Nevertheless, the approach reported here is based on direct quantity losses as in Ghosh's original position. This is justified by the fact that we are dealing with a downturn, and only over a short-term period.

ESTIMATION OF DIRECT IMPACTS

We rely on two data sets to estimate the direct impacts at the Port of New Orleans: WISERTrade data from the World Institute for Strategic Economic Research and data from the Waterborne Commerce of the United States (WCUS). We study two periods accounted for in the WISERTrade data: the months January 2003 through March 2006 and annual data from 2003. We examined 2003 WCUS data for the Louisiana Customs District, excluding the Mississippi river area, to eliminate possible double-counting of shipments.

The WCUS does not provide the monthly data needed for time-series forecasts of domestic imports and exports. As a result, we use annual data to calculate the ratios of short tons domestic (D_i) imports and exports to short tons of foreign (F_i) imports and exports of short tons for 2003. From WCUS:

$$R_i^E = \frac{D_i^E}{F_i^E}, \tag{10.1}$$

where R_i^E is the ratio of short tons of domestic exports to short tons of foreign export short tons in 2003 for each USC sector $i = 1$ to 29. The superscript E refers to exports.

Similarly, R_i^I is defined as the ratio of domestic imports to foreign imports,

$$R_i^I = \frac{D_i^I}{F_i^I},$$ (10.2)

where the superscript I refers to imports.

See Appendix Table 10A.1 for the estimated values R_i^E and R_i^I.

We convert the WISERTrade monthly export and import data into USC sectors to create W_i^E and W_i^I, respectively. Then, by multiplying the ratios R_i^E and R_i^I by W_i^E and W_i^I, respectively, the domestic values can be calculated as ${}^tDW_i^E$ and ${}^tDW_i^I$ for the months t = January 2003 through March 2006, a total of 39 observations,

$$^tDW_i^E = {}^tW_i^E \times R_i^E,$$ (10.3)

$$^tDW_i^I = {}^tW_i^I \times R_i^I.$$ (10.4)

Equations (10.3) and (10.4) provide two additional sets of monthly data, providing a total of four monthly data sets (${}^tW_i^E$, ${}^tDW_i^E$, ${}^tW_i^I$, and ${}^tDW_i^I$). The series of white coordinates in Figures 10.1a and 10.1b give $\Sigma_i^tW_i^E$ and $\Sigma_i^tDW_i^E$, respectively; and in Figures 10.2a and 10.2b give $\Sigma_i^tW_i^I$ and $\Sigma_i^tDW_i^I$. These monthly seaborne trade series for the given period are consistent with reference data for the Customs District of Louisiana.

We want to obtain differences between actual and forecast trade volumes. Observations for the period prior to Hurricane Katrina, January 2003 through July 2005, were used to forecast values for the August 2005 through March 2006 period based on the Holt-Winters times-series approach, and adjusting regular seasonal effects. See equation (10.5):

$$^tY_i = ({}^t\beta_{1i} + {}^t\beta_{2i})\,{}^tm_i + {}^t\varepsilon_i,$$ (10.5)

where, Y denotes each of the series W^E, DW^E, W^I, and DW^I, the terms ${}^t\beta_{1i}$ and ${}^t\beta_{2i}$ are trend coefficients, corresponding to the smoothed average values of the trend and the smoothed linear trend, respectively, for month t, and the terms ${}^tm_i$ are monthly seasonal adjustments parameters for each USC sector i.

In equation (10.5), parameters are updated according to equations (10.6a–c)

$$^t\beta_{1i} = \omega_{1i}\frac{{}^tY_i}{{}^{t-1}m_i} + (1-\omega_{1i})({}^{t-1}\beta_{1i} + {}^{t-1}\beta_{2i}),$$ (10.6a)

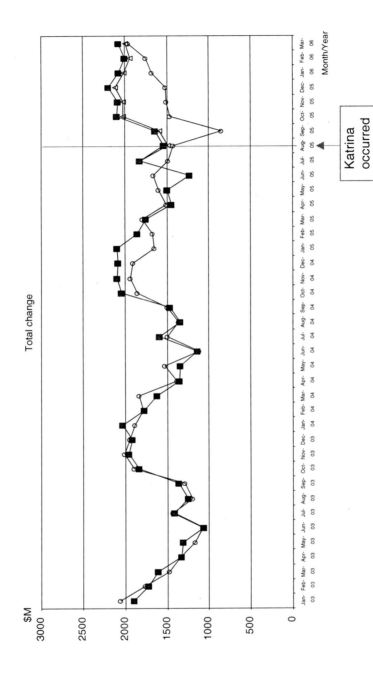

Figure 10.1a   Actual, estimated and forecast foreign exports by month

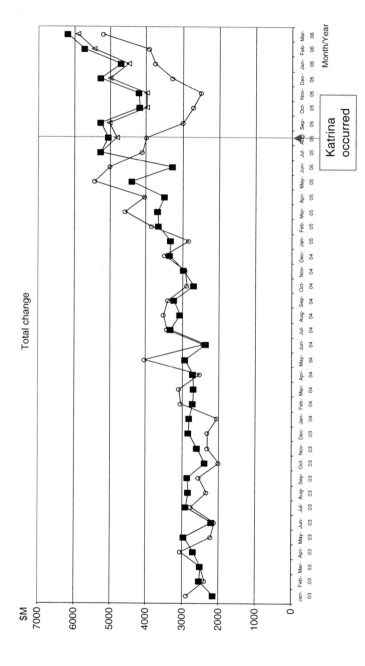

Note: The Holt-Winters time series method is used with monthly seasonal factors to account for seasonal effects. The black squares are estimates and forecasts. The white triangles are forecasts minus 10 percent. The white circles are observed values.

Figure 10.1b   Actual, estimated and forecast domestic exports by month

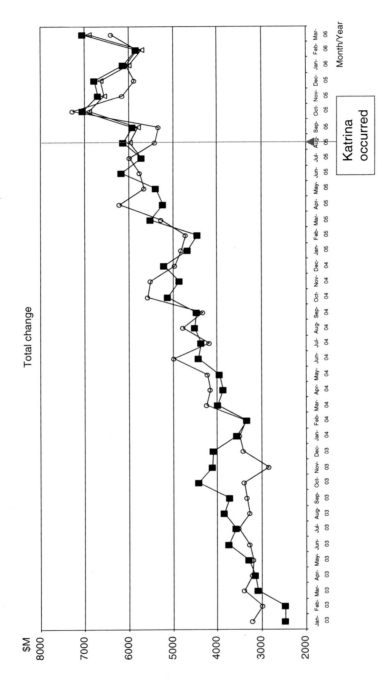

Total change

Katrina
occurred

Month/Year

$M

Figure 10.2a   Actual, estimated, and forecast foreign imports by month

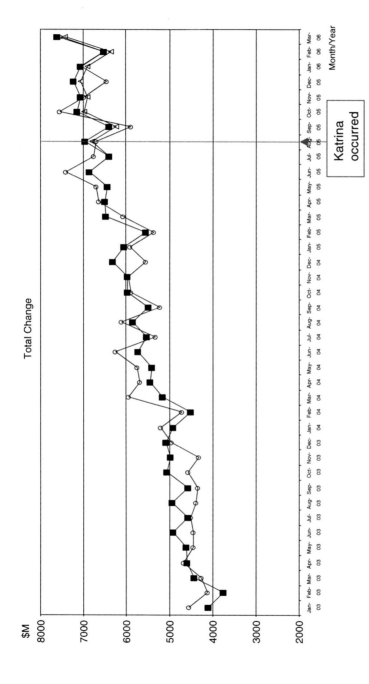

Note: The Holt-Winters time series method is used with monthly seasonal factors to account for seasonal effects. The black squares are estimates and forecasts. The white triangles are forecasts minus 10 percent. The white circles are observed values.

Figure 10.2b   Actual, estimated, and forecast domestic imports by month

157

$$'\beta_{2i} = \omega_{2i}\,('\beta_{1i} - {}^{t-1}\beta_{1i}) + (1 - \omega_{2i})\,{}^{t-1}\beta_{2i}, \text{ and} \qquad (10.6b)$$

$$'m_i = \omega_{3i}\frac{'Y_i}{'\beta_i} + (1 - \omega_{3i})^{t-1}m_i \qquad\qquad (10.6c)$$

where, ω_{ki} ($k = 1$ to 3), is a smoothing weight. For ω_{1i} and ω_{2i}, the weights are $(1 - 0.8^{1/\lambda})$, where $\lambda = 2$, and hence, $\omega_{1i} = \omega_{2i} = 0.10557$. The smoothing weight ω_{3i} is fixed at 0.25.

As shown in equations (10.6a–c), this method allows estimated coefficients to change gradually over the months, based on observations in the previous month and exponentially declining weights. The estimated and forecast trends summed across the USC sectors are shown in the Figures 10.1a, 10.1b, 10.2a and 10.2b as a series of black coordinates. Trends estimated in the period before Katrina produce forecasts for the period after Katrina. Given the actual and forecast data for four monthly trade series in the period after Katrina, the direct impacts of the hurricane are calculated as the differences between the actual and forecast trade values. These direct impacts are divided into two periods, the end of 2005 (August to December) and the first quarter of 2006 (January to March). Direct impacts for the four series thus produce a total of eight cases.

Defining the direct impacts to be Ψ, these are calculated as,

$$\Psi_i^Y = {}^{2005}\Psi_i^Y + {}^{2006}_{Q1}\Psi_i^Y \qquad\qquad (10.7a)$$

$$= \sum_{t=Sep.2005}^{Dec.2005} {}^t\Psi_i^Y + \sum_{t=Jan.2006}^{Mar2006} {}^t\Psi_i^Y \qquad\qquad (10.7b)$$

where the direct impact on foreign exports ${}^t\Psi_i^{FE} = {}^t\hat{W}_i^E - {}^tW_i^E,$

the direct impacts on foreign imports ${}^t\Psi_i^{FI} = {}^t\hat{W}_i^I - {}^tW_i^I,$

the direct impacts on domestic exports ${}^t\Psi_i^{DE} = D\hat{W}_i^E - {}^tDW_i^E,$ and

the direct impacts on domestic imports ${}^t\Psi_i^{DI} = D\hat{W}_i^E - {}^tDW_i^E.$

According to equations (10.7a–b), total direct impacts on the USC sectors, $\Sigma_i{}^t\Psi_i^{FE}, \Sigma_i{}^t\Psi_i^{DE}, \Sigma_i{}^t\Psi_i^{FI},$ and $\Sigma_i{}^t\Psi_i^{DI}$ are gaps denoting the differences between the forecast and observed series in Figures 10.1a and b and 10.2a and b. In the interests of conservatism, we reduced the forecast series by 10 percent prior to calculating direct impacts. See Appendix Tables 10A.2a and b and 10A.3a and b for the direct impacts ${}^t\Psi_i^{FE}, {}^t\Psi_i^{FI}, {}^t\Psi_i^{DE},$ and ${}^t\Psi_i^{DI}$ by USC sector; $t \geq$ August 2005.

NIEMO RESULTS

Tables 10.1a and 10.1b summarize our estimates on a state-by-state basis. Sectoral results are not shown. Eight kinds of results are displayed because there are two time periods, 2005 post-Katrina and the first quarter of 2006; results for foreign and domestic trade; as well as results for export losses and import losses. As in our previous work, the export loss impacts are handled via the demand-side NIEMO; this time, however, the import loss impacts are handled via the new supply-side NIEMO.

As the previous figures (and the Appendix tables) show, foreign seaborne exports from the customs district have slowly recovered, whereas domestic exports are only slowly recovering. Foreign imports have not recovered, whereas domestic import losses are less severe.

The four demand-side cases reveal aggregate US multipliers in the range of 2.2–2.4. The four supply-side multipliers are smaller (1.8–1.9). In fact the rightmost two columns cannot be described in terms of multipliers because there are net indirect effect gains in several of the states, leading to a positive sum for the nation as a whole.

In terms of distributions, for export losses, the effects in Louisiana are typically near three-quarters of the total effect. Beyond Louisiana, Texas and California show the largest indirect impacts. The maps of the states in Figures 10.3 through 10.8 contrast the geographic effects of import versus export losses. Aside from the two states, California and Texas, that are most affected, import loss impacts are more widely felt than export loss effects. Figure 10.9 sums total losses due to impacts on all imports and all exports. This combines results from the supply-side and demand-side versions of NIEMO, and should be considered provisional. There may be some double-counting of losses.

CONCLUSIONS

The allocation of Department of Homeland Security (DHS) funds to localities has been a contentious issue ever since the Department was created. National support for port protection comes from knowledge of how each port's activities impact the various states. In this research, we have shown that we can estimate state-by-state effects, not simply for the export losses involved but also for the import losses. As these are seen to recover at different paces, this added capability of NIEMO is useful.

Modeling efforts such as the one used in this research are complex and rely on a large number of assumptions. Input–output approaches underestimate the capacity of the market to implement substitutes. On the other

Table 10.1a State-by-state losses due to port closure, Customs District of Louisiana

State	Foreign export losses: 2005 ($m)		Foreign export losses: 1st quarter, 2006 ($m)		Domestic export losses: 2005 ($m)		Domestic export losses: 1st quarter, 2006 ($m)	
	Direct impacts	Total impacts	Direct impacts	Total impacts	Direct impacts	Total impacts	Direct impacts	Total impacts
AL	0.00	−26.74	0.00	−11.14	0.00	−95.56	0.00	−52.92
AK	0.00	−0.58	0.00	−0.24	0.00	−3.50	0.00	−2.42
AZ	0.00	−5.05	0.00	−1.97	0.00	−22.44	0.00	−11.36
AR	0.00	−19.31	0.00	−7.91	0.00	−72.50	0.00	−42.37
CA	0.00	−194.98	0.00	−52.66	0.00	−775.66	0.00	−234.75
CO	0.00	−6.23	0.00	−2.55	0.00	−24.26	0.00	−12.13
CT	0.00	−5.85	0.00	−2.24	0.00	−21.35	0.00	−10.69
DE	0.00	−1.39	0.00	−0.50	0.00	−3.99	0.00	−1.86
DC	0.00	−0.81	0.00	−0.48	0.00	−2.47	0.00	−1.61
FL	0.00	−45.97	0.00	−13.75	0.00	−126.55	0.00	−57.18
GA	0.00	−22.88	0.00	−9.05	0.00	−99.13	0.00	−58.08
HI	0.00	−0.40	0.00	−0.15	0.00	−1.28	0.00	−0.64
ID	0.00	−2.58	0.00	−1.16	0.00	−5.47	0.00	−2.68
IL	0.00	−33.41	0.00	−9.97	0.00	−107.56	0.00	−61.84
IN	0.00	−32.09	0.00	−12.76	0.00	−178.84	0.00	−107.46
IA	0.00	−7.35	0.00	−2.41	0.00	−20.83	0.00	−10.60
KS	0.00	−7.07	0.00	−2.52	0.00	−19.49	0.00	−9.02
KY	0.00	−13.17	0.00	−5.15	0.00	−51.31	0.00	−27.08
LA	−2765.72	−4803.42	−756.38	−1280.25	−8562.90	−14842.32	−3762.09	−6252.87
ME	0.00	−1.32	0.00	−0.53	0.00	−4.29	0.00	−2.08
MD	0.00	−4.97	0.00	−1.91	0.00	−17.42	0.00	−8.95
MA	0.00	−5.33	0.00	−2.10	0.00	−18.13	0.00	−8.80
MI	0.00	−26.39	0.00	−9.52	0.00	−148.02	0.00	−76.02
MN	0.00	−13.42	0.00	−4.17	0.00	−40.74	0.00	−20.22

MS	0.00	-45.99	0.00	-13.06	0.00	-112.93	0.00	-50.75
MO	0.00	-12.67	0.00	-4.54	0.00	-39.21	0.00	-19.95
MT	0.00	-1.43	0.00	-0.50	0.00	-5.61	0.00	-2.56
NE	0.00	-3.30	0.00	-1.32	0.00	-7.69	0.00	-3.71
NV	0.00	-1.00	0.00	-0.37	0.00	-3.36	0.00	-1.52
NH	0.00	-2.45	0.00	-0.95	0.00	-8.68	0.00	-4.22
NJ	0.00	-15.20	0.00	-5.35	0.00	-45.04	0.00	-23.44
NM	0.00	-2.70	0.00	-1.06	0.00	-11.45	0.00	-5.79
NY	0.00	-17.46	0.00	-6.69	0.00	-73.08	0.00	-39.96
NC	0.00	-21.78	0.00	-9.99	0.00	-58.18	0.00	-27.46
ND	0.00	-1.50	0.00	-0.34	0.00	-2.51	0.00	-1.26
OH	0.00	-27.84	0.00	-10.98	0.00	-125.46	0.00	-65.47
OK	0.00	-15.15	0.00	-5.70	0.00	-58.79	0.00	-27.14
OR	0.00	-4.49	0.00	-1.67	0.00	-17.72	0.00	-9.94
PA	0.00	-36.47	0.00	-15.44	0.00	-177.82	0.00	-109.64
RI	0.00	-1.73	0.00	-0.70	0.00	-6.14	0.00	-3.09
SC	0.00	9.07	0.00	-3.37	0.00	-30.36	0.00	-14.88
SD	0.00	-1.41	0.00	-0.64	0.00	-3.16	0.00	-1.69
TN	0.00	-21.54	0.00	-7.97	0.00	-88.28	0.00	-50.49
TX	0.00	-321.05	0.00	-104.58	0.00	-1233.92	0.00	-493.91
UT	0.00	-3.75	0.00	-1.33	0.00	-14.81	0.00	-6.69
VM	0.00	-0.79	0.00	-0.28	0.00	-2.38	0.00	-1.16
VA	0.00	-12.94	0.00	-4.70	0.00	-33.83	0.00	-16.45
WA	0.00	-9.71	0.00	-4.04	0.00	-33.80	0.00	-16.30
WV	0.00	-6.53	0.00	-2.31	0.00	-19.17	0.00	-10.23
WI	0.00	-20.06	0.00	-6.61	0.00	-57.35	0.00	-27.09
WY	0.00	-1.51	0.00	-0.48	0.00	-4.54	0.00	-1.84
US Total	-2765.72	-5900.26	-756.38	-1650.02	-8562.90	-18978.37	-3762.09	-8110.29
Rest of World	0.00	-332.43	0.00	-109.36	0.00	-1381.73	0.00	-544.87
World Total	2765.72	6232.69	-756.38	-1759.36	-8562.90	-20360.10	-3762.09	-8655.10

Table 10.1a (continued)

State	Foreign import losses: 2005 ($m)		Foreign import losses: 1st quarter, 2005 ($m)		Domestic import losses: 1st quarter, 2005 ($m)		Domestic import losses: 1st quarter, 2006 ($m)	
	Direct impacts	Total impacts	Direct impacts	Total impacts	Direct impacts	Total impacts	Direct impacts	Total impacts
AL	0.00	-22.29	0.00	-10.21	0.00	3.00	0.00	3.26
AK	0.00	-13.66	0.00	-4.41	0.00	-1.28	0.00	1.29
AZ	0.00	-20.77	0.00	-7.47	0.00	-0.12	0.00	2.54
AR	0.00	-25.11	0.00	-8.66	0.00	-8.18	0.00	-1.53
CA	0.00	-78.71	0.00	-33.65	0.00	0.82	0.00	5.09
CO	0.00	-7.57	0.00	-2.93	0.00	1.28	0.00	1.30
CT	0.00	-11.43	0.00	-4.05	0.00	-2.31	0.00	-0.22
DE	0.00	-1.85	0.00	-0.63	0.00	-0.11	0.00	0.13
DC	0.00	-4.01	0.00	-1.26	0.00	-0.80	0.00	0.32
FL	0.00	-44.52	0.00	-15.29	0.00	-31.80	0.00	-7.27
GA	0.00	-22.91	0.00	-8.10	0.00	-9.55	0.00	-2.50
HI	0.00	-1.91	0.00	-0.66	0.00	-0.51	0.00	-0.03
ID	0.00	-1.75	0.00	-0.60	0.00	-0.22	0.00	0.09
IL	0.00	-23.17	0.00	-9.91	0.00	10.55	0.00	7.82
IN	0.00	-23.85	0.00	-8.68	0.00	-1.26	0.00	1.71
IA	0.00	-13.40	0.00	-4.43	0.00	-2.41	0.00	0.50
KS	0.00	-11.63	0.00	-4.12	0.00	-2.88	0.00	-0.19
KY	0.00	-11.31	0.00	-3.95	0.00	-6.30	0.00	-1.47
LA	-2 533.44	-3 348.23	-729.76	-971.04	-1 226.04	-1 409.39	-101.23	-10.58
ME	0.00	-2.84	0.00	-0.94	0.00	-1.05	0.00	-0.17
MD	0.00	-5.85	0.00	-2.03	0.00	-0.16	0.00	0.54
MA	0.00	-16.07	0.00	-5.36	0.00	-4.09	0.00	-0.51
MI	0.00	-26.55	0.00	-9.06	0.00	-1.18	0.00	2.04
MN	0.00	-17.78	0.00	-6.18	0.00	-2.30	0.00	0.38

MS	0.00	−72.19	0.00	−25.26	0.00	−32.85	0.00	−10.54
MO	0.00	−10.09	0.00	−4.42	0.00	1.43	0.00	1.68
MT	0.00	−0.84	0.00	−0.31	0.00	0.21	0.00	0.24
NE	0.00	−4.20	0.00	−1.51	0.00	−0.58	0.00	0.24
NV	0.00	−3.80	0.00	−1.35	0.00	−0.47	0.00	0.13
NH	0.00	−3.64	0.00	−1.21	0.00	−0.89	0.00	−0.07
NJ	0.00	−13.85	0.00	−4.88	0.00	−2.20	0.00	0.17
NM	0.00	−5.00	0.00	−1.71	0.00	−1.12	0.00	−0.09
NY	0.00	−5.14	0.00	−3.88	0.00	8.80	0.00	5.04
NC	0.00	24.76	0.00	3.01	0.00	37.28	0.00	13.91
ND	0.00	−2.63	0.00	−0.97	0.00	−0.17	0.00	0.21
OH	0.00	−40.58	0.00	−13.81	0.00	−4.07	0.00	2.32
OK	0.00	−18.07	0.00	−7.02	0.00	4.01	0.00	3.19
OR	0.00	−5.55	0.00	−2.07	0.00	0.30	0.00	0.54
PA	0.00	−33.28	0.00	−11.81	0.00	−8.65	0.00	−1.12
RI	0.00	−2.88	0.00	−1.06	0.00	−0.43	0.00	0.02
SC	0.00	−9.23	0.00	0.47	0.00	16.77	0.00	6.34
SD	0.00	−1.34	0.00	−0.45	0.00	−0.18	0.00	0.06
TN	0.00	−18.60	0.00	−5.90	0.00	−7.15	0.00	−1.07
TX	0.00	−205.22	0.00	−75.21	0.00	−32.48	0.00	1.90
UT	0.00	−8.71	0.00	−3.00	0.00	−1.28	0.00	0.04
VM	0.00	−0.53	0.00	−0.15	0.00	−0.24	0.00	−0.01
VA	0.00	−42.45	0.00	−13.73	0.00	−5.66	0.00	3.48
WA	0.00	−10.62	0.00	−3.94	0.00	0.01	0.00	0.93
WV	0.00	1.94	0.00	−0.10	0.00	4.11	0.00	1.60
WI	0.00	−19.87	0.00	−6.78	0.00	−3.30	0.00	0.36
WY	0.00	−1.99	0.00	−0.68	0.00	−0.58	0.00	−0.06
US Total	−2533.44	−4252.31	−729.76	−1311.35	−1226.04	−1499.67	−101.23	31.99
Rest of World	0.00	−229.37	0.00	−77.53	0.00	−40.58	0.00	11.81
World Total	−2533.44	−4481.68	−729.76	−1388.88	−1226.04	−1540.25	−101.23	43.80

Table 10.1b Proportional state-by-state losses due to port closure, Customs District of Louisiana

State	Foreign export losses: 2005		Foreign export losses: 1st quarter, 2006		Domestic export losses: 2005		Domestic export losses: 1st quarter, 2006	
	Direct impacts	Total impacts	Direct impacts	Total impacts	Direct impacts	Total impacts	Direct impacts	Total impacts
AL	0.00%	0.43%	0.00%	0.63%	0.00%	0.47%	0.00%	0.61%
AK	0.00%	0.01%	0.00%	0.01%	0.00%	0.02%	0.00%	0.03%
AZ	0.00%	0.08%	0.00%	0.11%	0.00%	0.11%	0.00%	0.13%
AR	0.00%	0.31%	0.00%	0.45%	0.00%	0.36%	0.00%	0.49%
CA	0.00%	3.13%	0.00%	2.99%	0.00%	3.81%	0.00%	2.71%
CO	0.00%	0.10%	0.00%	0.14%	0.00%	0.12%	0.00%	0.14%
CT	0.00%	0.09%	0.00%	0.13%	0.00%	0.10%	0.00%	0.12%
DE	0.00%	0.02%	0.00%	0.03%	0.00%	0.02%	0.00%	0.02%
DC	0.00%	0.01%	0.00%	0.03%	0.00%	0.01%	0.00%	0.02%
FL	0.00%	0.74%	0.00%	0.78%	0.00%	0.62%	0.00%	0.66%
GA	0.00%	0.37%	0.00%	0.51%	0.00%	0.49%	0.00%	0.67%
HI	0.00%	0.01%	0.00%	0.01%	0.00%	0.01%	0.00%	0.01%
ID	0.00%	0.04%	0.00%	0.07%	0.00%	0.03%	0.00%	0.03%
IL	0.00%	0.54%	0.00%	0.57%	0.00%	0.53%	0.00%	0.71%
IN	0.00%	0.51%	0.00%	0.73%	0.00%	0.88%	0.00%	1.24%
IA	0.00%	0.12%	0.00%	0.14%	0.00%	0.10%	0.00%	0.12%
KS	0.00%	0.11%	0.00%	0.14%	0.00%	0.10%	0.00%	0.10%
KY	0.00%	0.21%	0.00%	0.29%	0.00%	0.25%	0.00%	0.31%
LA	100.00%	77.07%	100.00%	72.77%	100.00%	72.90%	100.00%	72.24%
ME	0.00%	0.02%	0.00%	0.03%	0.00%	0.02%	0.00%	0.02%
MD	0.00%	0.08%	0.00%	0.11%	0.00%	0.09%	0.00%	0.10%
MA	0.00%	0.09%	0.00%	0.12%	0.00%	0.09%	0.00%	0.10%
MI	0.00%	0.42%	0.00%	0.54%	0.00%	0.73%	0.00%	0.88%
MN	0.00%	0.22%	0.00%	0.24%	0.00%	0.20%	0.00%	0.23%
MS	0.00%	0.74%	0.00%	0.74%	0.00%	0.55%	0.00%	0.59%

MO	0.00%	0.20%	0.00%	0.26%	0.00%	0.19%	0.00%	0.23%
MT	0.00%	0.02%	0.00%	0.03%	0.00%	0.03%	0.00%	0.03%
NE	0.00%	0.05%	0.00%	0.08%	0.00%	0.04%	0.00%	0.04%
NV	0.00%	0.02%	0.00%	0.02%	0.00%	0.02%	0.00%	0.02%
NH	0.00%	0.04%	0.00%	0.05%	0.00%	0.04%	0.00%	0.05%
NJ	0.00%	0.24%	0.00%	0.30%	0.00%	0.22%	0.00%	0.27%
NM	0.00%	0.04%	0.00%	0.06%	0.00%	0.06%	0.00%	0.07%
NY	0.00%	0.28%	0.00%	0.38%	0.00%	0.36%	0.00%	0.46%
NC	0.00%	0.35%	0.00%	0.57%	0.00%	0.29%	0.00%	0.32%
ND	0.00%	0.02%	0.00%	0.02%	0.00%	0.01%	0.00%	0.01%
OH	0.00%	0.45%	0.00%	0.62%	0.00%	0.62%	0.00%	0.76%
OK	0.00%	0.24%	0.00%	0.32%	0.00%	0.29%	0.00%	0.31%
OR	0.00%	0.07%	0.00%	0.09%	0.00%	0.09%	0.00%	0.11%
PA	0.00%	0.59%	0.00%	0.88%	0.00%	0.87%	0.00%	1.27%
RI	0.00%	0.03%	0.00%	0.04%	0.00%	0.03%	0.00%	0.04%
SC	0.00%	0.15%	0.00%	0.19%	0.00%	0.15%	0.00%	0.17%
SD	0.00%	0.02%	0.00%	0.04%	0.00%	0.02%	0.00%	0.02%
TN	0.00%	0.35%	0.00%	0.45%	0.00%	0.43%	0.00%	0.58%
TX	0.00%	5.15%	0.00%	5.94%	0.00%	6.06%	0.00%	5.71%
UT	0.00%	0.06%	0.00%	0.08%	0.00%	0.07%	0.00%	0.08%
VM	0.00%	0.01%	0.00%	0.02%	0.00%	0.01%	0.00%	0.01%
VA	0.00%	0.21%	0.00%	0.27%	0.00%	0.17%	0.00%	0.19%
WA	0.00%	0.16%	0.00%	0.23%	0.00%	0.17%	0.00%	0.19%
WV	0.00%	0.10%	0.00%	0.13%	0.00%	0.09%	0.00%	0.12%
WI	0.00%	0.32%	0.00%	0.38%	0.00%	0.28%	0.00%	0.31%
WY	0.00%	0.02%	0.00%	0.03%	0.00%	0.02%	0.00%	0.02%
US Total	100.00%	94.67%	100.00%	93.78%	100.00%	93.21%	100.00%	93.70%
Rest of World	0.00%	51.36%	0.00%	6.22%	0.00%	6.19%	0.00%	0.30%
World Total	100.00%	100.00%	100.00%	100.00%	100.00%	100.00%	100.00%	100.00%

Table 10.1b (continued)

State	Foreign import losses: 2005		Foreign import losses: 1st quarter, 2006		Domestic import losses: 2005		Domestic import losses: 1st quarter, 2006	
	Direct impacts	Total impacts	Direct impacts	Total impacts	Direct impacts	Total impacts	Direct impacts	Total impacts
AL	0.00%	0.50%	0.00%	0.74%	0.00%	−0.19%	0.00%	7.44%
AK	0.00%	0.30%	0.00%	0.32%	0.00%	0.08%	0.00%	2.95%
AZ	0.00%	0.46%	0.00%	0.54%	0.00%	0.01%	0.00%	5.79%
AR	0.00%	0.56%	0.00%	0.62%	0.00%	0.53%	0.00%	−3.49%
CA	0.00%	1.76%	0.00%	2.42%	0.00%	−0.05%	0.00%	11.62%
CO	0.00%	0.17%	0.00%	0.21%	0.00%	−0.08%	0.00%	2.97%
CT	0.00%	0.26%	0.00%	0.29%	0.00%	0.15%	0.00%	−0.51%
DE	0.00%	0.04%	0.00%	0.05%	0.00%	0.01%	0.00%	0.30%
DC	0.00%	0.09%	0.00%	0.09%	0.00%	0.05%	0.00%	0.73%
FL	0.00%	0.99%	0.00%	1.10%	0.00%	2.06%	0.00%	−16.60%
GA	0.00%	0.51%	0.00%	0.58%	0.00%	0.62%	0.00%	−5.71%
HI	0.00%	0.04%	0.00%	0.05%	0.00%	0.03%	0.00%	−0.07%
ID	0.00%	0.04%	0.00%	0.04%	0.00%	0.01%	0.00%	0.21%
IL	0.00%	0.52%	0.00%	0.71%	0.00%	−0.68%	0.00%	17.85%
IN	0.00%	0.53%	0.00%	0.62%	0.00%	0.08%	0.00%	3.91%
IA	0.00%	0.30%	0.00%	0.32%	0.00%	0.16%	0.00%	1.14%
KS	0.00%	0.26%	0.00%	0.30%	0.00%	0.19%	0.00%	−0.44%
KY	0.00%	0.25%	0.00%	0.28%	0.00%	0.41%	0.00%	−3.35%
LA	100.00%	74.71%	100.00%	69.92%	100.00%	91.50%	100.00%	−24.15%
ME	0.00%	0.06%	0.00%	0.07%	0.00%	0.07%	0.00%	−0.39%
MD	0.00%	0.13%	0.00%	0.15%	0.00%	0.01%	0.00%	1.24%
MA	0.00%	0.36%	0.00%	0.39%	0.00%	0.27%	0.00%	−1.15%
MI	0.00%	0.59%	0.00%	0.65%	0.00%	0.08%	0.00%	4.67%
MN	0.00%	0.40%	0.00%	0.45%	0.00%	0.15%	0.00%	0.86%

	1	2	3	4	5	6	7	8
MS	−24.06%	0.00%	2.13%	0.00%	1.82%	0.00%	1.61%	0.00%
MO	3.83%	0.00%	−0.09%	0.00%	0.32%	0.00%	0.23%	0.00%
MT	0.55%	0.00%	−0.01%	0.00%	0.02%	0.00%	0.02%	0.00%
NE	0.55%	0.00%	0.04%	0.00%	0.11%	0.00%	0.09%	0.00%
NV	0.30%	0.00%	0.03%	0.00%	0.10%	0.00%	0.08%	0.00%
NH	−0.17%	0.00%	0.06%	0.00%	0.09%	0.00%	0.08%	0.00%
NJ	0.38%	0.00%	0.14%	0.00%	0.35%	0.00%	0.31%	0.00%
NM	−0.20%	0.00%	0.07%	0.00%	0.12%	0.00%	0.11%	0.00%
NY	11.51%	0.00%	−0.57%	0.00%	0.28%	0.00%	0.11%	0.00%
NC	31.77%	0.00%	−2.42%	0.00%	−0.22%	0.00%	−0.55%	0.00%
ND	0.47%	0.00%	0.01%	0.00%	0.07%	0.00%	0.06%	0.00%
OH	5.29%	0.00%	0.26%	0.00%	0.99%	0.00%	0.91%	0.00%
OK	7.28%	0.00%	−0.26%	0.00%	0.51%	0.00%	0.40%	0.00%
OR	1.24%	0.00%	−0.02%	0.00%	0.15%	0.00%	0.12%	0.00%
PA	−2.56%	0.00%	0.56%	0.00%	0.85%	0.00%	0.74%	0.00%
RI	0.05%	0.00%	0.03%	0.00%	0.08%	0.00%	0.06%	0.00%
SC	14.48%	0.00%	−1.09%	0.00%	−0.03%	0.00%	−0.21%	0.00%
SD	0.14%	0.00%	0.01%	0.00%	0.03%	0.00%	0.03%	0.00%
TN	−2.45%	0.00%	0.46%	0.00%	0.42%	0.00%	0.41%	0.00%
TX	4.35%	0.00%	2.11%	0.00%	5.42%	0.00%	4.58%	0.00%
UT	0.08%	0.00%	0.08%	0.00%	0.22%	0.00%	0.19%	0.00%
VM	−0.03%	0.00%	0.02%	0.00%	0.01%	0.00%	0.01%	0.00%
VA	7.94%	0.00%	0.37%	0.00%	0.99%	0.00%	0.95%	0.00%
WA	2.13%	0.00%	0.00%	0.00%	0.28%	0.00%	0.24%	0.00%
WV	3.66%	0.00%	−0.27%	0.00%	0.01%	0.00%	−0.04%	0.00%
WI	0.82%	0.00%	0.21%	0.00%	0.49%	0.00%	0.44%	0.00%
WY	−0.14%	0.00%	0.04%	0.00%	0.05%	0.00%	0.04%	0.00%
US Total	73.04%	100.00%	97.37%	100.00%	94.42%	100.00%	94.88%	100.00%
Rest of World	26.96%	0.00%	2.63%	0.00%	5.58%	0.00%	5.12%	0.00%
World Total	100.00%	100.00%	100.00%	100.00%	100.00%	100.00%	100.00%	100.00%

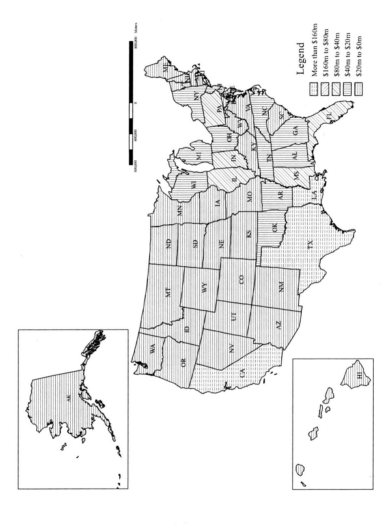

Legend

More than $160m
$160m to $80m
$80m to $40m
$40m to $20m
$20m to $0m

Figure 10.3   The state-by-state total economic impacts of reductions in foreign exports through the Customs District of Louisiana following Hurricanes Katrina and Rita, 1 August 2005 through 31 March 2006

168

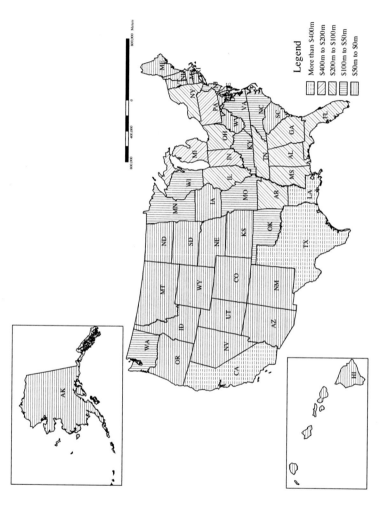

Figure 10.4   The state-by-state total economic impacts of reductions in domestic exports through the Customs District of Louisiana following Hurricanes Katrina and Rita, 1 August 2005 through 31 March 2006

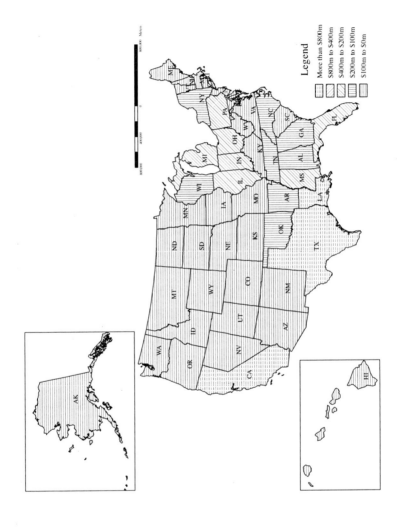

Figure 10.5   The state-by-state total economic impacts of reductions in foreign and domestic exports through the Customs District of Louisiana following Hurricanes Katrina and Rita, 1 August 2005 through 31 March 2006

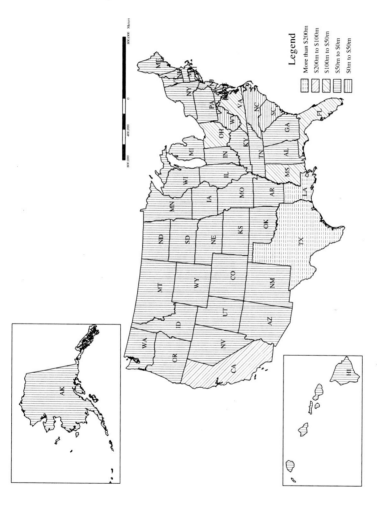

Figure 10.6 The state-by-state total economic impacts of reductions in foreign imports through the Customs District of Louisiana following Hurricanes Katrina and Rita, 1 August 2005 through 31 March 2006

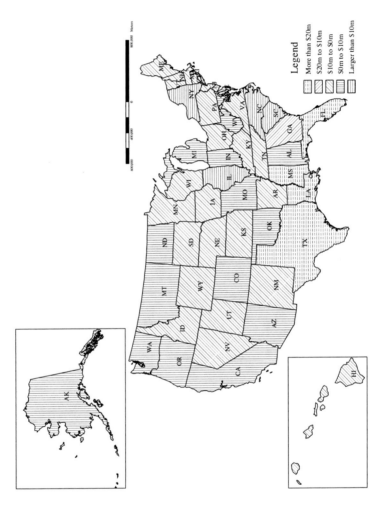

Legend

More than $20m
$20m to $10m
$10m to $0m
$0m to $10m
Larger than $10m

Figure 10.7   The state-by-state total economic impacts of reductions in domestic imports through the Customs District of Louisiana following Hurricanes Katrina and Rita, 1 August 2005 through 31 March 2006

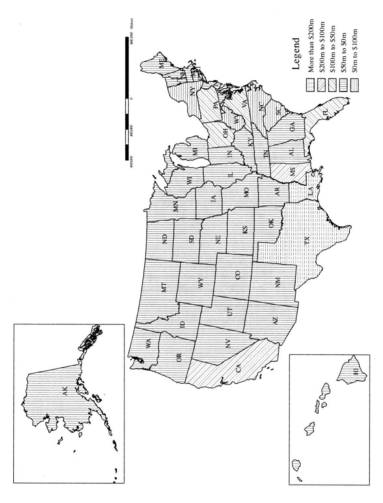

Figure 10.8   The state-by-state total economic impacts of reductions in foreign and domestic imports through the Customs District of Louisiana following Hurricanes Katrina and Rita, 1 August 2005 through 31 March 2006

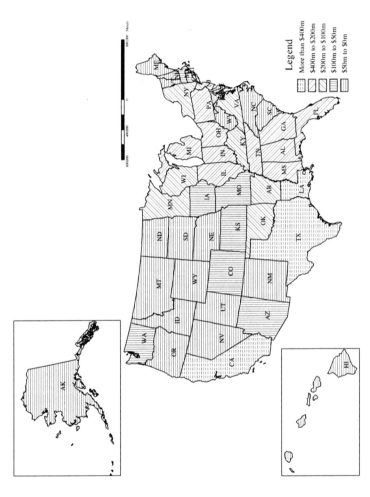

Figure 10.9 *The state-by-state total economic impacts of reductions in foreign and domestic imports and exports through the Customs District of Louisiana following Hurricanes Katrina and Rita, 1 August 2005 through 31 March 2006*

hand, they make it possible to estimate very detailed impacts. We have not included induced effects in our estimates because we are wary of including interstate effects of labor movements, which are probably minor, and we are happy to be conservative in our applications of the model.

Our interest in the effects of Katrina stems from the fact that it is not a hypothetical event, and we can develop estimates on better data than from the application of hypothetical scenarios. As gross state product data become available for 2005 and 2006 from the Bureau of Economic Analysis we will be able to get a better sense of the accuracy of NIEMO and also on where and how to improve it.

NOTE

* This research was supported by the United States Department of Homeland Security through the Center for Risk and Economic Analysis of Terrorism Events (CREATE) under grant number N00014-05-0630. However, any opinions, findings, and conclusions or recommendations in this document are those of the authors and do not necessarily reflect views of the United States Department of Homeland Security.

REFERENCES

Bon, R. (2001), 'Comparative stability analysis of demand-side and supply-side input–output models: toward an index of economic "Maturity"', in M.L. Lahr, and E. Dietzenbacher (eds), *Input–Output Analysis: Frontiers and Extensions*, New York: Palgrave.

Bon, R. and T. Yashiro (1995), 'Comparative stability analysis of demand-side and supply-side input–output models: the case of Japan, 1960–90', *Applied Economics Letters*, **3**, 349–54.

Chenery, H.B. (1953), 'Regional analysis', in H.B. Chenery, P.G. Clark and V.C. Pinna (eds), *The Structure and Growth of the Italian Economy*, Rome: US Mutual Security Agency, pp. 98–139.

Dietzenbacher, E. (1997), 'In vindication of the Ghosh model: a reinterpretation as a price model', *Journal of Regional Science*, **37**, 629–51.

Dietzenbacher, E. (2002), 'Interregional multipliers: looking backward, looking forward', *Regional Studies*, **36** (2), 125–36.

Ghosh, A. (1958), 'Input–output approach to an allocative system', *Economica*, **25**, 58–64.

Gordon, P., J.E. Moore II, H.W. Richardson and Q. Pan (2005), 'The economic impact of a terrorist attack on the twin ports of Los Angeles–Long Beach', in H.W. Richardson, P. Gordon and J.E. Moore II (eds), *The Economic Impacts of Terrorist Attacks*, Cheltenham, UK and Northampton, MA, USA: Edward Elgar, pp. 262–82.

Gruver, G.W. (1989), 'On the plausibility of the supply-driven input–output model: a theoretical basis for input-coefficient change', *Journal of Regional Science*, **29**, pp. 441–50.

Horwich, G. (2000), 'Economic lessons of the Kobe earthquake', *Economic Development and Cultural Change*, **48** (3), 521–42.

Isard, W. (1951), 'Interregional and regional input–output analysis: a model of a space economy', *Review of Economics and Statistics*, **33**, 318–28.

Jack Faucett Associates (1983), '*The multiregional input–output Accounts, 1977: introduction and summary, Vol. I (final report)*', prepared for the US Department of Health and Human Services, Washington, DC.

Jackson, R.W., W.R. Schwarm, Y. Okuyama and S. Islam (2006), 'A method for constructing commodity by industry flow matrices', *Annals of Regional Science*, **40** (4), 909–20.

Leontief, W. (1936), 'Quantitative input and output relations in the economic system of the United States', *Review of Economic Statistics*, **18** (3), 105–25.

Leontief, W. (1941), *The Structure of the American Economy, 1919–1929: An Empirical of Equilibrium Analysis*, Cambridge, MA: Harvard University Press.

Leontief, W. (1976), *The Structure of the American Economy, 1919–1939: An Empirical of Equilibrium Analysis*, White Plains, NY: International Arts and Science Press.

Lindall, S., D. Olsen and G. Alward (2005), 'Deriving multi-regional models using the IMPLAN national trade flows model', Paper presented at the 2005 MCRSA/SRSA Annual Meeting, 7–9 April, Arlington, VA.

Moses, L.N. (1955), 'The stability of interregional trading patterns and input–output analysis', *American Economic Review*, **45**, 803–32.

Oosterhaven, J. (1988), 'On the plausibility of the supply-driven input–output model', *Journal of Regional Science*, **28**, 203–17.

Oosterhaven, J. (1989), 'The supply-driven input–output model: a new interpretation but still implausible', *Journal of Regional Science*, **29**, 459–65.

Oosterhaven, J. (1996), 'Leontief versus Ghoshian price and quantity models', *Southern Economic Journal*, **62** (3), 750–59.

Park, J.Y. (2006a), 'The economic impacts of a dirty-bomb attack on the Los Angeles and Long Beach Port: applying supply-driven NIEMO', *Journal of Homeland Security and Emergency Management*, **5** (1), Article 21, available at http://www.bepress.com/jhsem/vol5/iss1/21.

Park, J.Y. (2006b), 'Estimation of state-by-state trade flows of service sectors applying geographical weighted regression', paper presented at the North American Meetings of the Regional Science Association International 53rd Annual Conference.

Park, J.Y. and P. Gordon (2005), 'An evaluation of input–output aggregation error using a new MRIO model', paper presented at North American Meetings of the Regional Science Association International 52nd Annual Conference, Las Vegas, NV, 10–12 November.

Park, J.Y., P. Gordon, J.E. Moore II and H.W. Richardson (2004), 'Construction of a US multiregional input output model using IMPLAN', paper presented at 2004 National IMPLAN User's Conference, Shepherdstown, WV, 6–8 October.

Park, J.Y., C.K. Park and S.J. Nam (2006), 'The state-by-state effects of mad cow disease using a new MRIO model', paper presented at the 2006 American Agricultural Economic Association (AAEA) Annual Meeting, Long Beach, CA, 23–26 July.

Park, J.Y., P. Gordon, J.E. Moore II, H.W. Richardson and L. Wang (2007), 'Simulating the state-by-state effects of terrorist attacks on three major US ports: applying NIEMO (National Interstate Economic Model)', in H.W. Richardson,

P. Gordon and J.E. Moore II (eds), *The Economic Costs and Consequences of Terrorism*, Cheltenham, UK and Northampton, MA, USA: Edward Elgar.

Polenske (1980), 'The US multiregional input–output accounts and model', DC Health, Lexington, MA.

Richardson, H.W., P. Gordon, J.E. Moore II, S.J. Kim, J.Y. Park and Q. Pan (2007a), 'Tourism and terrorism: the national and interregional economic impacts of attacks on major US theme parks', in H.W. Richardson, P. Gordon and J.E. Moore II (eds), *The Economic Costs and Consequences of Terrorism*, Cheltenham, UK and Northampton, MA, USA: Edward Elgar.

Richardson, H.W., P. Gordon and J.E. Moore II (2007b), *The Economic Costs and Consequences of Terrorism*, Cheltenham, UK and Northampton, MA, USA: Edward Elgar.

Rose, A. and T. Allison (1989), 'On the plausibility of the supply-driven input–output model: empirical evidence on joint stability', *Journal of Regional Science*, **29**, 451–8.

Wang, E.C. (1997), 'Structural change and industrial policy in Taiwan, 1966–91: an extended input–output analysis', *Asian Economic Journal*, **11** (2), 187–206.

APPENDIX

Table 10A.1   Ratio of domestic exports to foreign exports by USC sector

USC sector	D_i^E (1000 ton)	F_i^E (1000 ton)	R_i^E	D_i^I (1000 ton)	F_i^I (1000 ton)	R_i^I
USC 01	2	478	0.0045	1	14	0.0357
USC 02	1 268	29 965	0.0423	15 387	542	28.4015
USC 03	179	1 660	0.1080	468	111	4.2192
USC 04	13	653	0.0199	26	12	2.1655
USC 05	1 379	1 681	0.8206	1 799	832	2.1612
USC 06	5	35	0.1429	0	79	0.0000
USC 07	0	1	0.0000	0	0	0.0000
USC 08	10 575	1046	10.1110	14 217	3 913	3.6332
USC 09	6 125	115	53.4308	91	6 035	0.0150
USC 10	47 220	6 605	7.1487	34 142	48 867	0.6987
USC 11	7 883	3 063	2.5733	4 347	3 547	1.2256
USC 12	30	103	0.2922	0	71	0.0035
USC 13	4 478	231	19.3853	1 565	2 861	0.5470
USC 14	155	527	0.2937	279	136	2.0527
USC 15	5	619	0.0088	1	652	0.0008
USC 16	88	269	0.3262	73	284	0.2568
USC 17	8	574	0.0132	22	73	0.2930
USC 18	58	14	4.0922	51	10	5.0370
USC 19	57	79	0.7243	39	40	0.9760
USC 20	661	307	2.1565	1 908	1 063	1.7956
USC 21	4 796	685	7.0045	717	5 062	0.1417
USC 22	1 932	122	15.8371	963	456	2.1117
USC 23	1 371	105	12.9998	534	99	5.3867
USC 24	88	188	0.4657	40	164	0.2459
USC 25	1 425	79	17.9565	572	100	5.7041
USC 26	56	43	1.2832	40	8	5.3667
USC 27	55	56	0.9888	40	63	0.6407
USC 28	55	5	10.5092	39	5	7.8000
USC 29	8 459	1 167	7.2470	2 055	1 400	1.4673
Total	98 426	50 475	1.9500	79 414	76 499	1.0381

Notes: $R_i^{E(I)}$ = ratio of short tons of domestic exports (imports) to short tons of foreign exports (imports) in 2003 for each USC sector i = 1 to 29.
$D_i^{E(I)}$ = short tons of domestic exports (imports).
$F_i^{E(I)}$ = short tons of foreign exports (imports).

Table 10A.2a Direct impacts due to loss of foreign exports (Ψ_i^{FE}) by USC sector (m)

Month	Aug 05	Sep 05	Oct 05	Nov 05	Dec 05	2005	Jan 06	Feb 06	Mar 06	Q1 2006
USC1	11.4	−11.2	−21.8	−25.4	−18.4	−65.4	−18.3	−6.6	−13.6	−38.5
USC2	−16.4	−233.7	−164.4	−212.9	−295.9	−923.3	−64.7	58.6	112	105.9
USC3	6.4	9.8	22.4	−32.2	2.9	9.3	−33.5	34.8	−23.4	−22.1
USC4	11	6.4	0.6	3.2	−3.1	18.1	0	15.6	9	24.6
USC5	12	−12.6	−1.6	11.9	18.1	27.8	−4.6	6.6	9.9	11.9
USC6	−0.6	−0.5	0.4	−0.6	−0.1	−1.4	−0.3	−0.7	−0.3	−1.3
USC7	0	0	0	0	0	0	0	0	0	0
USC8	0.1	−0.3	−0.3	0.3	−0.2	−0.4	0	0.2	0.3	0.5
USC9	−12.4	4	−4.1	−16.1	1.6	−27	−2.5	−9.2	−11.1	−22.8
USC10	−21	−129.5	−67.4	51	−131	−297.9	−34.7	−92.6	80	−47.3
USC11	−40.6	−187.5	−163.9	−217.2	−32.4	−641.6	−157.9	−99.1	−28.5	−285.5
USC12	0	−0.6	−0.5	−0.3	0	−1.4	−0.8	−0.3	−0.1	−1.2
USC13	14.9	−0.1	−0.2	−4.1	−1	9.5	1.9	0.8	3.5	6.2
USC14	−2.3	−25.4	−22.2	−4.8	−23.3	−78	−0.7	−8.8	−13.2	−22.7
USC15	5.5	−55.2	−55.4	−27.5	−17.1	−149.7	−29.7	−33.3	−18.2	−81.2

Table 10A.2a (continued)

Month	Aug 05	Sep 05	Oct 05	Nov 05	Dec 05	2005	Jan 06	Feb 06	Mar 06	Q1 2006
USC16	0.8	−5.5	−3.4	0	−2.1	−10.2	0.2	−2.5	−1	−3.3
USC17	−0.2	−17.4	−18.4	−15.4	−6.6	−58	−10.1	−10.8	−8.3	−29.2
USC18	−1.1	−15.4	−0.7	−0.7	−2.8	−20.7	−0.7	−1.4	−0.5	−2.6
USC19	−3.1	−13.4	−11.9	−14.8	−15.9	−59.1	−11.6	−16.5	−17.4	−45.5
USC20	0.4	−4.4	−2.7	−4.5	−11.1	−22.3	−0.6	−2.1	−7	−9.7
USC21	−6.3	−2	−6.8	2.6	−3.8	−16.3	−3	−11.3	−11.7	−26
USC22	−0.4	−3.5	4.1	−2.7	−1.4	−3.9	−8.1	−3.6	−5.3	−17
USC23	−14.4	−55.5	−10.2	−30.2	−14.5	−124.8	−3.8	−0.1	−9.1	−13
USC24	−1.2	−5	−3.6	−3.6	2.3	−11.1	−3.1	−0.5	−0.5	−4.1
USC25	−3	−11.6	−3.4	−3	−29.5	−50.5	−3.7	−1.6	−22.9	−28.2
USC26	−26.4	−16.1	−82.4	−28.9	−66.3	−220.1	−12.7	−34.9	−126.4	−174
USC27	−4.9	−2	2.5	0	−2	−6.4	0.4	−6.5	−0.8	−6.9
USC28	−0.4	−0.8	−0.7	−0.4	−0.5	−2.8	−0.8	−0.5	−0.6	−1.9
USC29	−17.3	6.8	−2.9	−5	−19.3	−37.7	15.8	−25.5	−12	−21.7
Total	−109.6	−782.2	−618.9	−581.6	−673.5	−2765.8	−387.6	−251.6	−117.2	−756.4

Table 10A.2b Direct impacts due to loss of domestic exports (Ψ_i^{DE}) by USC sector (m)

Month	Aug 05	Sep 05	Oct 05	Nov 05	Dec 05	2005	Jan 06	Feb 06	Mar 06	Q1 2006
USC1	0.1	0	-0.1	-0.1	-0.1	-0.2	-0.1	0	-0.1	-0.2
USC2	-0.7	-9.9	-7	-9	-12.5	-39.1	-2.7	2.5	4.7	4.5
USC3	0.7	1.1	2.4	-3.5	0.3	1	-3.6	3.8	-2.5	-2.3
USC4	0.2	0.1	0	0.1	-0.1	0.3	0	0.3	0.2	0.5
USC5	9.8	-10.4	-1.3	9.7	14.9	22.7	-3.8	5.4	8.1	9.7
USC6	-0.1	-0.1	0.1	-0.1	0	-0.2	0	-0.1	0	-0.1
USC7	0	0	0	0	0	0	0	0	0	0
USC8	1.1	-3.3	-3.2	2.7	-1.7	-4.4	-0.5	1.6	2.7	3.8
USC9	-661.2	214.3	-219.4	-860.3	87.3	-1 439.3	-134	-491.7	-592	-1 217.7
USC10	-150.4	-925.7	-481.7	364.9	-936.4	-2 129.3	-247.8	-661.9	572.2	-337.5
USC11	-104.5	-482.6	-421.8	-559	-83.5	-1 651.4	-406.3	-254.9	-73.3	-734.5
USC12	0	-0.2	-0.1	-0.1	0	-0.4	-0.2	-0.1	-0.1	-0.3
USC13	288.3	-2.3	-3.5	-80.4	-18.6	183.5	36.7	16.4	68.8	121.9
USC14	-0.7	-7.5	-6.5	-1.4	-6.8	-22.9	-0.2	-2.6	-3.9	-6.7
USC15	0	-0.5	-0.5	-0.2	-0.2	-1.4	-0.3	-0.3	-0.2	-0.8

Table 10A.2b (continued)

Month	Aug 05	Sep 05	Oct 05	Nov 05	Dec 05	2005	Jan 06	Feb 06	Mar 06	Q1 2006
USC16	0.2	−1.8	−1.1	0	−0.7	−3.4	0.1	−0.8	−0.3	−1
USC17	0	−0.2	−0.2	−0.2	−0.1	−0.7	−0.1	−0.1	−0.1	−0.3
USC18	−4.4	−63.1	−2.7	−2.8	−11.6	−84.6	−2.9	−5.6	−1.9	−10.4
USC19	−2.3	−9.7	−8.7	−10.7	−11.5	−42.9	−8.4	−11.9	−12.6	−32.9
USC20	0.8	−9.5	−5.9	−9.8	−24	−48.4	−1.2	−4.5	−15.1	−20.8
USC21	−44.1	−14.3	−47.6	18.3	−26.4	−114.1	−21.2	−78.9	−82.2	−182.3
USC22	−5.9	−55.3	64.9	−42	−21.4	−59.7	−127.7	−56.9	−84.6	−269.2
USC23	−186.8	−721.6	−133.1	−392.5	−188.5	−1 622.5	−49.1	−1.1	−117.7	−167.9
USC24	−0.6	−2.3	−1.7	−1.7	1.1	−5.2	−1.4	−0.3	−0.2	−1.9
USC25	−53	−208.3	−61.8	−54.6	−530.4	−908.1	−67	−29.5	−411.9	−508.4
USC26	−33.8	−20.7	−105.7	−37.1	−85.1	−282.4	−16.4	−44.7	−162.2	−223.3
USC27	−4.8	−2	2.5	0	−2	−6.3	0.4	−6.5	−0.8	−6.9
USC28	−4.3	−8.3	−7.3	−4.7	−4.8	−29.4	−8	−4.9	−6	−18.9
USC29	−125.6	49.5	−21.3	−36.6	−140.1	−274.1	114.2	−184.8	−87	−157.6
Total	−1 081.9	−2 294.4	−1 472.2	−1 711.3	−2 003.1	−8 562.9	−951.7	−1 812.3	−998.1	−3 762.1

Table 10A.3a Direct impacts due to loss of foreign imports ($^t\Psi_i^{FI}$) by USC sector ($m)

Month	Aug 05	Sep 05	Oct 05	Nov 05	Dec 05	2005	Jan 06	Feb 06	Mar 06	Q1 2006
USC1	-1.3	-3.4	-0.9	-0.4	-1.7	-7.7	-1.4	-2.2	-1.6	-5.2
USC2	-9.9	-14.8	-8.4	-9.2	-3.5	-45.8	-7.8	-7.1	-7.4	-22.3
USC3	-5.2	-11.6	-2.3	-3.7	-1.8	-24.6	3.5	-1.4	-3.3	-1.2
USC4	0.1	0	-0.3	-0.4	0	-0.6	0.1	-0.4	0.2	-0.1
USC5	-23.9	-6.3	-0.9	4.6	12.3	-14.2	12.3	-3.4	19.2	28.1
USC6	5.5	3.6	-5.9	-8.5	-1.3	-6.6	1.7	8.9	1.9	12.5
USC7	0.1	0.1	-0.2	-0.1	-0.2	-0.3	-0.1	-0.1	-0.1	-0.3
USC8	-13.9	7.2	3.9	8.2	-5.5	-0.1	-2.5	-3.9	-3.4	-9.8
USC9	-9	-35.1	0.8	3.7	-2.6	-42.2	-16.1	-9.9	-28.3	-54.3
USC10	-181.8	296.1	664	277.3	-281.2	774.4	311.7	233.7	-372.2	173.2
USC11	-35.5	-69.8	50.3	92.4	20.7	58.1	42.5	33.6	23	99.1
USC12	48.5	75.7	26.9	-8.6	10	152.5	7.6	2.3	32.1	42
USC13	-38	-10	-38.8	-77.5	-57.6	-221.9	16.5	21.7	-49	-10.8
USC14	-6.5	-8.9	-15.4	-9.3	-10.2	-50.3	-3.2	-12.2	-2.2	-17.6
USC15	5.8	-50.2	-11.8	-0.4	6.5	-50.1	35.8	-9.9	24.4	50.3

Table 10A.3a (continued)

Month	Aug 05	Sep 05	Oct 05	Nov 05	Dec 05	2005	Jan 06	Feb 06	Mar 06	Q1 2006
USC16	-20.1	-18.5	-20.8	-12.7	-17.1	-89.2	-10.2	3.2	-11.3	-18.3
USC17	0.3	-1.3	-5.5	-3.9	-3.4	-13.8	-3.4	-5.5	-3.1	-12
USC18	3.6	5.5	6.9	2.6	4	22.6	2.6	3.9	3.3	9.8
USC19	-13.7	-40.1	-39	-60.8	-53.6	-207.2	-38.1	-47.1	-50.4	-135.6
USC20	-8.5	-11.4	-4.5	-9.8	-7	-41.2	-6.4	-9.8	4.6	-11.6
USC21	-490.5	-711.2	-452	-729.5	-477.7	-2860.9	-410.8	-227	-385.2	-1023
USC22	-14.3	-24.1	21.4	-23.1	-2.1	-42.2	-8.8	-5	25.5	11.7
USC23	61.2	-15.2	-0.5	34.5	-25.7	54.3	5.6	11.2	102	118.8
USC24	45.6	15.4	19.7	17.3	17.2	115.2	1.9	11	58.1	71
USC25	29.5	45.1	60.9	13.3	-34	114.8	9.6	31.8	7.3	48.7
USC26	-0.3	-0.5	-4	-1.2	-2.1	-8.1	-1.3	-2.2	2	-1.5
USC27	-7.8	-0.1	-6.3	-2.8	2.4	-14.6	6.5	-3.8	-9.3	-6.6
USC28	-2	-15.4	-12.8	-12.3	-4.2	-46.7	-9.3	-10.9	-24.4	-44.6
USC29	-22.8	-6.2	-12.8	-25.5	29.8	-37.5	1.3	-13.9	-7.4	-20
Total	-704.6	-605.1	211.5	-545.8	-889.4	-2533.4	-60.2	-14.5	-655	-729.7

Table 10A.3b Direct impacts due to loss of domestic imports (Ψ_i^{DI}) by USC sector ($m)

Month	Aug 05	Sep 05	Oct 05	Nov 05	Dec 05	2005	Jan 06	Feb 06	Mar 06	Q1 of 2006
USC1	0	−0.1	0	0	−0.1	−0.2	−0.1	−0.1	−0.1	−0.3
USC2	−280.9	−419	−239.2	−260.3	−98	−1297.4	−222.1	−201.4	−209.3	−632.8
USC3	−21.8	−48.9	−9.9	−15.5	−7.6	−103.7	14.6	−5.8	−14	−5.2
USC4	0.2	−0.1	−0.7	−0.8	0.1	−1.3	0.2	−0.9	0.3	−0.4
USC5	−51.6	−13.5	−1.9	9.9	26.6	−30.5	26.6	−7.3	41.4	60.7
USC6	0	0	0	0	0	0	0	0	0	0
USC7	0	0	0	0	0	0	0	0	0	0
USC8	−50.4	26	14.2	29.7	−20.1	−0.6	−9.2	−14.1	−12.2	−35.5
USC9	−0.1	−0.5	0	0.1	0	−0.5	−0.2	−0.1	−0.4	−0.7
USC10	−127	206.9	463.9	193.7	−196.5	541	217.8	163.3	−260	121.1
USC11	−43.5	−85.5	61.6	113.3	25.4	71.3	52.1	41.2	28.2	121.5
USC12	0.2	0.3	0.1	0	0	0.6	0	0	0.1	0.1
USC13	−20.8	−5.5	−21.2	−42.4	−31.5	−121.4	9	11.9	−26.8	−5.9
USC14	−13.4	−18.3	−31.6	−19	−20.9	−103.2	−6.5	−25.1	−4.4	−36
USC15	0	0	0	0	0	0	0	0	0	0

185

Table 10A.3b (continued)

Month	Aug 05	Sep 05	Oct 05	Nov 05	Dec 05	2005	Jan 06	Feb 06	Mar 06	Q1 of 2006
USC16	-5.2	-4.7	-5.3	-3.3	-4.4	-22.9	-2.6	0.8	-2.9	-4.7
USC17	0.1	-0.4	-1.6	-1.2	-1	-4.1	-1	-1.6	-0.9	-3.5
USC18	17.9	27.9	34.8	13.3	20.1	114	13.3	19.6	16.4	49.3
USC19	-13.3	-39.1	-38	-59.3	-52.4	-202.1	-37.2	-46	-49.2	-132.4
USC20	-15.3	-20.5	-8.2	-17.6	-12.6	-74.2	-11.4	-17.6	8.2	-20.8
USC21	-69.5	-100.8	-64	-103.4	-67.7	-405.4	-58.2	-32.2	-54.6	-145
USC22	-30.3	-50.8	45.3	-48.8	-4.4	-89	-18.6	-10.7	53.8	24.5
USC23	329.7	-81.8	-3	185.9	-138.3	292.5	30.1	60.1	549.6	639.8
USC24	11.2	3.8	4.8	4.3	4.2	28.3	0.5	2.7	14.3	17.5
USC25	168.1	257.5	347.5	76	-194.1	655	54.5	181.2	41.6	277.3
USC26	-1.5	-2.5	-21.5	-6.5	-11.3	-43.3	-6.9	-11.6	10.6	-7.9
USC27	-5	-0.1	-4	-1.8	1.5	-9.4	4.1	-2.4	-5.9	-4.2
USC28	-15.9	-120.4	-99.8	-95.9	-32.6	-364.6	-72.7	-85.1	-190.5	-348.3
USC29	-33.4	-9.1	-18.8	-37.4	43.7	-55	2	-20.4	-10.8	-29.2
Total	-271.4	-499.3	403.5	-87	-771.8	-1226	-21.9	-1.8	-77.5	-101.2

11. Regional economic impacts of natural and man-made hazards: disrupting utility lifeline services to households

Adam Rose and Gbadebo Oladosu

INTRODUCTION

Nearly all of the literature on the regional economic impacts of disruptions to utility lifelines has focused on industrial and commercial customers, or business interruption. However, residential customers purchase 30–40 percent of electricity, gas and water services, and are therefore worthy of attention. Disruption of household activities is not just an inconvenience, but can affect health and safety at home and productivity on the job. Moreover, even the use of leisure time to boil water or to purchase fuel for a backup electric generator has a value, as does family tranquility.

Hurricane Katrina dramatized this situation even more. It generated indelible images of people suffering from dehydration and living in fear on dark streets. Not only were they cut off from major lines of communication because ordinary phone lines were down and cell phones could not be recharged, but their escape routes were limited, and protective service personnel had in many instances abandoned them. Although the death toll was not immense given the dimensions of the catastrophe, the human suffering was substantial (see, for example, Schlenger et al., 2006; Copeland, 2006).

On another front, terrorists, as well as their analysts, have typically focused on industrial, financial and utility targets in terms of the ability to cripple the economy by disrupting these institutions. However, there is every indication that an ultimate aim of terrorist activity is to destroy the morale of the general public. Contaminating the water supply or cutting off the electricity supply may have more immediate effects on disrupting the lives and comfort zone of the citizenry than would an attack on an industrial target.

To date, most of the studies of the household impacts of disasters, including terrorist attacks, have focused on social and psychological manifestations such as trauma in the short run, and afflictions such as alcoholism, mental illness and family dysfunction in the long run. There are several reasons why more attention has not been placed on economic impacts on individuals (other than potential job and income losses). First is a bias against including non-market activities because they are not thought to have value or because the value is difficult to measure. Second is the belief that these impacts are small in comparison to more conventional business interruption. Third is the limitation of economic impact models. Input–output (I–O) models are dominated by production-side considerations, with consumption being perfunctory and household behavior being non-existent. Computable general equilibrium (CGE) models do incorporate behavior, but the sophistication of the consumer (household) side typically lags behind that of the producer side of this approach (Rose, 2005a).

The situation, however, is that the value of household activities can be measured, impacts on them from extreme events are potentially large, and the ability to model them can be enhanced. A simple set of back-of-the-envelope calculations can demonstrate the relative importance of these impacts. Take for example a decree to boil water, affecting 1 million customers for seven days, due to contamination of the water supply. Estimates are that this situation will scare customers away from restaurants in the affected area, resulting in a $10 million loss of business revenue (though translating this lost revenue into value-added and adjusting for substitution effects of spending this money elsewhere in the economy reduces this figure to $2 million–$3 million). The increased demand for electricity to boil the water amounts to less than $1 million. However, even if we utilize conservative numbers of households spending only 15 minutes per day to boil and store water at an average wage rate (opportunity costs of leisure time) of $25 per hour, this results in an economic cost to them of over $40 million.

In this chapter, we report on the development of a capability to analyze, both theoretically and empirically, the impact of natural hazards and terrorist activities on households directly and through their interaction with the rest of the economy. This will be done with the use of household production functions (see, for example, Becker, 1965; Oladosu, 2000). It calls for the enhancement of the consumer components of the CGE framework to model household activities that combine market-based commodities with non-market based inputs (primarily household time). It will also include both pre-event protective behavior or mitigation (storing water, purchasing backup electric generators, and so on), and post-event recovery or resilience (conservation, substitution, and so on).

In addition to convenience costs and changes in household demand on the quantity of goods produced directly and indirectly, the effects on labor productivity represent an important feedback loop to the rest of the economy. We also comment on how Hurricane Katrina has identified even greater challenges in modeling household activities and how they can be addressed.

The analysis represents a generalization of economic impact modeling that has several additional useful applications. The specification of a household production function represents a way of incorporating the value of non-market goods into loss estimation. This includes not only the value of unpriced household services, but the value of time in general, as applicable to the large cost of transportation delays to commuters and shoppers (see, for example, Cho et al., 2001). It is also applicable to estimating the value of typically unpriced public infrastructure (for example, bridges, highways, airports) (see Shinozuka et al., 1998; Gordon et al., 2005). Finally, it should enable analysts to estimate the value of environmental goods and services (see Oladosu, 2000), which are gaining increased attention in the natural hazards literature (see, for example, Heinz Center, 2000) and which may be express targets of terrorist attacks, as in the case of national parks and watersheds, or that are typically devastated by coastal hazards, as in the case of estuaries and natural fisheries. The analysis can be extended to a broader range of targets that are part of the non-market economy, such as national monuments.

The improvement in the sophistication of the modeling of consumer behavior also facilitates the calculation of 'welfare measures' of impacts.[1] These microeconomic-based metrics are more widely used in economic policy analysis than are macroeconomic-based indicators such as output, income and employment. Output as a measure of sales, employment as a measure of economic vitality, and income as a measure of potential consumption and savings, are readily recognizable, and perhaps preferred indicators to businessmen, policymakers, and the general public, respectively. However, welfare measures are considered to reflect the value of resources more accurately. For example, recreation demand in a standard multi-sector loss estimation model is confined to the value of travel expenses, hotel and restaurant revenues, and where applicable, admission fees. This omits several aspects of consumer willingness to pay for access to sites such as national treasures, including both use and non-use values (see Brookshire, 2005). These additional values are much easier to model in a fully specified consumer component of a regional economic impact model. Their estimation is also more consistent with the use of welfare measures than with conventional macro impact indicators.

CONCEPTUAL FRAMEWORK

Our analytical framework extends the traditional focus of multi-sector macroeconomic models in a number of ways as illustrated in Figure 11.1. The boxes represent entities or objects, the lines represent flows, and the arrows show their direction. The conventional view of the economic system, presented in the center of Figure 11.1, includes two categories of entities – businesses and households, with two distinct functions – production and consumption. The circular flow of the economy is depicted by households' provision of factors of production (capital and labor) to businesses in exchange for factor payments (wages, dividends, interest, rents and royalties).

Multi-sector models add an important dimension to the standard representation, by acknowledging not only that final goods and services are transacted in the conventional manner, but that there are important intermediate goods (mainly industrial materials and business services) that are exchanged between firms.[2] The standard I–O analysis focuses on disruption of the flow of these intermediate goods on customers or suppliers in terms of multiplier or general equilibrium effects. Other dimensions of these broader effects relate to household income payments and spending, but not to household operations. Standard I–O models would analyze a hurricane or terrorist attack in terms of a demand or production capacity change, with the focus on the supply response. CGE models offer a more balanced approach to the two components through the formal interaction of supply and demand.

The two-stage depiction of production provides a segue to the expanded representation of households in our model. Rather than households simply being consumers of products, it shows their representation as producers of various household goods and services (transportation, cooking, recreation), which combine purchased inputs, household-produced (intermediate) inputs and time. There may be no explicit payment in the use of household labor, but it does have an opportunity cost as given by the market wage rate. Likewise, the use of capital within the household has an opportunity cost reflected by the market interest rate. Thus, the household represents a sub-model of circular flows as well.

The interruption of household activity by a terrorist attack can lead to a number of impacts:

- Reduction in household leisure time in relation to protective measures and adaptive responses.
- Reallocation of time to a different mix of household productive activities, including previously unnecessary activities.
- Reduction in quantity or quality of household-produced goods and services.

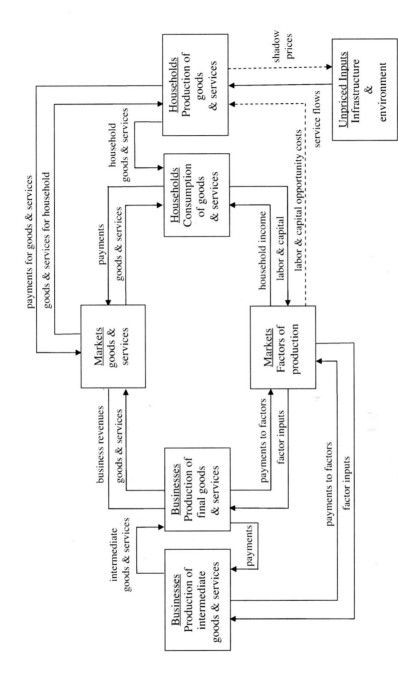

Figure 11.1   Expanded circular flow of the economy

- Reduction in the quantity or quality of factors of production that households provide to themselves and to businesses.

The latter manifests itself in ways such as reduced productivity of labor in the business production function, thereby representing another important link between the two sectors. This is in addition to the more standard links in I–O or CGE models of reduced household spending on goods and services produced by the business sector.

The framework also shows the role of non-market inputs in the household sector. The flows are primarily unidirectional in that their services are provided, but there is not always a return flow of payments (as distinguished by the dashed line). An increasing number of user charges (for example, road tolls, entrance fees, pollution emission permits) are, however, improving this situation. Even with the lack of explicit prices, however, it is possible to determine the value of these inputs through the calculation of 'shadow', or efficient, prices in a CGE model.[3]

THE HOUSEHOLD PRODUCTION FUNCTION TO MODELING CONSUMER BEHAVIOR

Most macro models use rather simple formulations on the consumer side in comparison to the production side. The former are typically specified in terms of linear expenditure systems or Cobb–Douglas utility functions of market-based goods and services, in contrast to nested constant elasticity of substitution functions or flexible functional forms (translog or generalized Leontief) and the inclusion of non-market commodities. As such, conventional models omit important costs and offer limited input substitution possibilities, as well omitting important life functions relating to leisure, recreation and maintaining day-to-day household operations.

General Considerations

Household production function (HPF) theory dates back to the original work by Becker (1965) and Muth (1966) and has been characterized by several important advances in recent years (see, for example, Pollak and Wales, 1992; Chiappori, 1997; Kerkhofs and Kooreman, 2003). The theory posits that households do not necessarily derive utility directly from purchases of market goods but rather from the use of market goods in combination with time and non-market (typically) public goods to produce commodities for consumption. Household-produced commodities may also be substitutes for some market goods. For example, HPF theory would not view automobiles as direct sources of utility, but as inputs into a household production process, along with fuel, time, and so on, to produce private transportation. Private transportation (the household-produced commodity) in turn is a substitute for market transportation such as public bus systems and commercial airlines.

HPF theory implies a reformulation of the consumer's problem as a multi-stage optimization process. Consumers maximize utility (from commodities) subject to a budget constraint in the first stage, and minimize the cost of producing those commodities in subsequent stages. A complete accounting of market goods inputs into commodities is contained in a household input–output table or technology matrix.

Pollack and Wachter (1975) pointed out that this formulation of the consumer's problem complicates the process of solving for demand of (market) goods and services, since household production functions are characterized by joint production and non-constant returns to scale. Moreover, commodity prices of contributed inputs and commodity outputs are endogenous to each household in relation to household marginal costs or shadow prices. Time inputs are difficult to allocate to commodities because of multitasking and the onerous nature of record keeping. Finally, actual inputs and outputs of household commodities are non-observable. However, Barnett (1977) and Mendelsohn (1984) pointed out that these issues are empirical (that is, measurement and econometric) rather than theoretical. Given this situation, empirical use of the HPF framework often involves simplifying assumptions about the nature of household production 'technologies'.

Hori (1975) established a classification of four types of household production technologies:

1. Each market good is used in the production of only one household commodity.
2. Each market good is divided up into multiple, but distinct, production of household commodities.
3. Each market good contributes to the production of multiple household commodities indivisibly.
4. A mixture of (2) and (3).

A generalized, or complete, implementation of the HPF framework would involve Type (4) technologies but implies dealing with the full set of empirical issues highlighted above. By invoking the assumptions of linear homogeneity (constant returns to scale) and non-jointness that are commonly made in the specification of computable general equilibrium (CGE) models, difficulties in implementing the HPF framework in a CGE context are considerably reduced (Oladosu, 2000). However, these assumptions mean that only Type (2) technologies, with Type (1) as a special case, can be modeled.

A prototype of the consumer expenditure system based on Type (2) technologies can be specified following Oladosu (2000):

$$U_h = U_h(Z_{h1}, Z_{h2}) \quad \text{household utility functions} \tag{11.1}$$

$$Z_{hc} = Z_{hc}(X_{hca}, X_{hcb}) \quad \text{household production functions} \tag{11.2}$$

$$X_{hi} = \Sigma_c X_{hci} \quad \text{purchase of market goods by households} \tag{11.3}$$

$$I_h = \Sigma_i P_i \Sigma_c X_{hci} = \Sigma_c P_{hc} Z_{hc} \quad \text{household budget constraint} \tag{11.4}$$

where:

U_h = household h's utility $h = 1 \ldots s$
Z_{hc} = household h's production of commodity c $c = 1 \ldots m$
X_{hci} = input of market good i into commodity c in household h $i = 1$
       $\ldots n$
X_{hi} = the observable total purchase of market good i by household h
P_i = market price of good i
P_{hc} = household h's marginal cost of commodity c

Although the assumptions of linear homogeneity and non-jointness reduce the difficulty of implementing the HPF in a CGE model, an additional issue arises with respect to uniqueness of its solution. This uniqueness requirement can be boiled down to ensuring that the endogenously determined household technology matrix (that is, household I–O table) has a non-zero determinant. A necessary but insufficient condition for this is that the number of market goods inputs into the production functions must be greater than or equal to the number of household-produced commodities. The uniqueness issue can be eliminated by having strictly Type (1) household technologies. The resulting expenditure system would be akin to use of consistent budget aggregates in traditional consumer expenditure systems, which also require the assumptions of constant returns to scale and non-jointness (the latter is implied by definition). This additional restriction would imply the following prototype formulation of what might be termed a 'budget aggregate' approach:

$$U_h = U_h(Z_{h1}, Z_{h2}) \quad \text{household utility functions} \tag{11.1}$$

$$Z_{hc} = Z_{hc}(X_{hc}) \quad \text{household production functions} \tag{11.5}$$

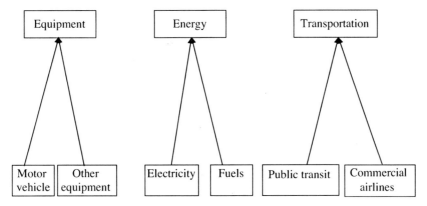

Figure 11.2   Type (1) Household technology formulation

$$I_h = \Sigma_i P_i X_{hi} = \Sigma_c P_{hc} Z_{hc} \quad \text{household budget constraint} \qquad (11.6)$$

where X_{hC} now represents the vector of market goods making up the budget aggregate Z_{hc} from the set of market goods ($i = 1. . .n$) that are purchased by household h. These budget aggregates are strictly not commodities as in the HPF framework, since no attention is paid to how households employ the market goods in 'production technologies'. However, depending on the area of application, the difference between Type (1) and Type (2) technology specifications may be minimized by including a finer level of detail in the set of market goods that serve as inputs into Type (1) technologies. Figures 11.2 and 11.3 illustrate the difference between the Type (1) and the Type (2) technology formulations.

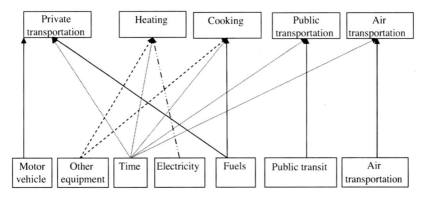

Figure 11.3   Type (2) Household technology formulation

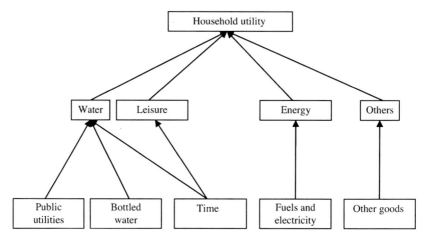

Figure 11.4 Household formulation of boil water decree (Type 1/Type 2 hybrid)

Sample Application: Assessing the Household Economic Effects of a Boil Water Notice

A 'boil water notice' instructs household about a way to destroy contaminants in publicly supplied water. This requirement could potentially affect a household's use of water in several ways:

1. Allocating time and energy to boil publicly supplied water for drinking and other purposes.
2. Substituting other types of water for publicly provided water, including self-supplied ground and surface water, and bottled water.
3. Substituting other goods for goods using public water, such as drinking beverages, eating out to avoid cooking (assuming restaurants are unaffected).
4. Reducing the total amount of water use (conservation).

A framework for capturing these effects is illustrated in Figure 11.4. Empirical implementation of this framework could employ the linear expenditure system (LES) to specify utility functions and constant elasticity of substitution (CES) functions to model household water production as below.

Linear expenditure system

$$V_h = \left(I_h - \sum_i \gamma_{i,h} P_{i,h} \right) \Pi_i \left(\frac{\alpha_i}{P_{i,h}} \right)^{\alpha_i} \tag{11.7}$$

where:

V_h = indirect utility function
I_h = household *h*'s income (including opportunity cost of leisure time)
P_{ih} = household *h*'s shadow price for commodity *i*
i = 1 . . . 4 (Water, Leisure, Energy, Others)
γ_{ih} = subsistence demand for commodity *i* by household *h*
α_{ih} = marginal expenditure shares for commodity *i* by household *h*

Differentiating equation (11.7) according to Roy's identity yields the household's demand for commodities.

Constant elasticity of substitution unit cost function
A constant elasticity of substitution (CES) function is used to determine the unit cost (price) of the commodities entering the utility function based on prices of component market goods. For the present exercise we let the Energy and Other Goods and Services component of variable P_{ih} be the same for all households, and calculated as some linear combination of market prices for Energy and Other Goods. The price for leisure can also be taken as equal to a given linear function of the household's wage rate. Thus, only the price for water is specified using a CES unit cost function:

$$P_{j,h} = a \left(\sum_k \lambda_k \left(b_k P_k \right)^{1-\sigma_{j,h}} \right)^{1/(1-\sigma_{j,h})}$$
(11.8)

where:
j = water
k = 1.2 (that is, public water and bottled water)
λ_j = CES share parameter
α_{ih} = household *h*'s substitution elasticity between public water and bottled water
σ = general CES shift parameter
b_j = CES shift parameter for individual goods
P_j = price of inputs

Differentiating the above with respect to each of the input prices yields endogenous input–output coefficients (or shares when differentiated in log form) from which the demand for inputs into the production of water can be derived. It would be difficult, and a stretch, to treat the input of time in this cost function in the same way as market goods, since its price is indirectly calculated from household wage rates. A solution to this would be to specify the unit cost function for households with only public and bottled water inputs as above, while relating public water to time through an auxiliary, fixed coefficient, equation. This fixed coefficient could then be changed to reflect normal and boil water scenarios, that is:

$$t_{publicwater,h} = tc * X_{publicwater,\ water,\ h} \qquad (11.9)$$

where:

$t_{publicwater,h}$ = time input into boiling publicly supplied water
$X_{publicwater,water,h}$ = public water input into household water use
tc = boiling time coefficient

The time involved in preparing public water for consumption would be subtracted from household leisure time, and its cost would be added to the overall cost of household water production. Since households have a fixed amount of time to allocate to labor, leisure and, in this case, boiling water, the effect of a 'boil water notice' would be transmitted through a household water shadow price change, substitution with bottled water, and reduction in leisure. The magnitude of these different effects would depend to a large extent on the elasticity of substitution between bottled water and public water.

Other improvisations of household responses to water disruptions would be variations on the above themes. For example, in the case of a shortage, use of rainfall or riverine water could be included as a new input variable subject to a constraint on availability (in a manner similar to inventories of stored water). A parameter similar to tc in this case, sc, would represent the cost of household conversion to usable water. In the case of electricity, where household storage is limited, the more likely response, in addition to conservation and substitution, is the use of a backup generator. This would involve adding another tier to the formulation above in which capital is combined with fuel and time to generate electricity. A further discussion of these responses and how they might be enhanced is presented in the following section.

Data Needs

The data needs for the above formulation in a CGE context include the following:

1. Data on public water and bottled water purchases to calibrate the cost function for water.
2. Estimates of labor–leisure allocation of time in the specified households, and applicable wage rates.

3. Estimates of purchased goods prices and time allocation for backup supplies.
4. Other economic data usually employed in CGE model formulation, primarily price elasticities for the LES expenditure system and the unit cost function.

Analogous data would be needed to analyze the response to electricity supply disruptions, including information on substitutes and the costs of purchasing and operating (including household time) backup generators. These data can be obtained from standard consumer expenditure surveys, utility company reports, and special analyses of how households cope with disasters (see Oladosu, 2006).

HOUSEHOLD RESILIENCE

The previous section provided an approach to modeling consumer behavior but with little attention to how crisis situations add an important additional dimension. Experience has revealed, and analysts have come to understand, that individuals react differently in a crisis. Rather than panic, the record shows a significant amount of favorable behavior exhibiting organizational skills and ingenuity (see, for example, Comfort, 1999; Rose, 2008). This behavior has more recently been characterized under the heading of 'resilience', which also applies to business, government and communities (economies) as a whole. The definition of resilience in a static mode is: the ability of a system to maintain function so as to reduce potential losses. In a dynamic context it is defined as: the rapidity at which a system returns to the desired state.[4] An operational measure of static resilience put forth by Rose (2004) is the potential maximum disruption from a given shock minus the actual disruption, all divided by the maximum disruption. In other words, resilience refers to the percentage avoidance of the maximum disruption.

Omission of resilience will lead to overestimation of negative impacts of disasters. Tierney (1997) and Rose and Lim (2002) estimated direct business resilience to a water supply disruption to be 77 and 90 percent, respectively, for the Northridge earthquake, and Rose et al. (2007) measured it at nearly 90 percent in response to a terrorist attack on the Los Angeles water system. There is also every reason to hypothesize that resilience would be large for households as well. Note also that just as the model developed in this chapter identifies a major factor whose omission might lead to an exaggeration of impacts, it also includes two factors whose omission will lead to underestimation of losses. The first is the major focus of the chapter: the inclusion of basic household impacts, including non-market costs. The second is general equilibrium effects of household activities, including those related to changes in the quantities of goods consumed and changes in the quantity of factors provided.

Resilience first appeared prominently in the field of ecology in the writing of Holling (1973) and has been refined by environmental and ecological economists. More recently, resilience has been utilized in the field of natural hazards (see, for example, Tierney, 1997; Rose et al., 1997; and the comprehensive formulation in Bruneau et al., 2003). The more specific formulation of the concept with a focus on economic dimensions is contained in Rose (2004) and Rose and Liao (2005). To date, most of the analysis of resilience has dealt with government, formal organizations and informal social arrangements, rather than the economic behavior of consumers.

In this chapter we extend the production-oriented analysis by Rose and Liao (2005) to the realm of the consumer. An important distinction between two types of resilience applies to this new focus as well:

- Inherent: the ability under normal circumstances (for example, individual households substituting other inputs for those disrupted by an external shock).
- Adaptive: ability in crisis situations due to ingenuity or extra effort (for example, increasing input substitution possibilities, conservation).

We begin by noting differences in the context between business and household resilience:

- Households have relatively lower competitive pressures than businesses.
- Households have less technical expertise.
- Households have fewer economies of scale (at least when households act individually).
- Households have fewer established procedures.
- Households have more flexibility in coping because of the absence of rigid managerial and production structures.

Second, we make an explicit distinction between mitigation and resilience. The former is considered to pertain primarily to pre-event measures to lower the probability of an adverse occurrence or its damage should it occur. The latter pertains to behavior during or after the disaster or disruption. Elsewhere, Rose (2005b) has couched disasters in economic terms by defining an 'economic disruption' in a manner more consistent

Table 11.1   *Adjustments to reduction in water quantity and quality to households*

Adjustment type	Household production function tier	Parameter change	Applicability (time trend)
1. Overall conservation of water use	CES Upper Tier	a	immediate to long run (increasing)
2. Decontamination of water	auxiliary equation	tc	immediate to long run (increasing)
3. Increased substitutability of other sources of water (e.g. substitute bottled water for public water)	CES Upper Tier	σ	short run to long run (increasing except inventories)
4. Backup supplies	auxiliary equation w/constraint	sc	immediate to long run (increasing)

with resilience, as the length of time until the economy (or at the micro level, the business or household activity) returns to its desired state. Thus, the disaster is not the short period of earthquake ground-shaking or terrorist explosion of a high-rise office building, but the longer period of response, recovery, remediation and reconstruction. Here, however, we will focus on what is referred to as the 'disaster response and recovery'.

More specifically, Table 11.1 lists a set of household responses to a natural or man-made disruption or contamination of water supplies. Next to each of the responses is the identification of the relevant household production function layer in which the behavior is reflected. The third column lists the specific production function parameter that captures inherent resilience, and the arrows indicate the direction of change in the parameter value associated with adaptive resilience. The final column indicates the time period of application and whether the adaptive resilience is likely to increase or decrease over time.

A more detailed summary of the responses is provided below:[5]

1. Conservation of water. This response is exemplified by ordinary reduction in the use of water due to foregoing consumption. Its presence is reflected in the unit cost function (equation 11.2 of the previous section) in the form of the a term. Conservation refers to a decrease in the value of this parameter, or an improvement in productive efficiency of water. This response can be undertaken immediately and remains

through the long run, and is likely to be able to increase over time, though at a decreasing rate, as more options are discovered or more experience is completed. Some of these practices might also become permanently ingrained into household operations.

2. Conservation of water-intensive inputs. In the household case, this would refer to examples such as decreasing cooking at home in general or cooking certain kinds of meals specifically, or eliminating use of swimming pools. This would again apply to equation (11.2) and to the *a* parameter, though to goods other than water (for example, the household production of the good 'swimming' in the latter case). The capability is similar to that of ordinary water conservation, though some of the other goods may be more seasonal in nature.

3. Decontamination of water. This refers to the production of a household commodity using a combination of priced inputs like electricity and time to boil water. It is basically an inherent option but it imposes extra costs and time on households. These costs would be reflected in increased electricity use, and change in the *tc* parameter for boiled water. Enhanced flexibility of households with respect to water use can be implemented by changing the substitution parameter, σ. It is applicable from the immediate to the long run, with adaptation increasing over time at a decreasing rate.

4. Increased substitutability of other inputs for water system deliveries. This response would be exemplified by purchasing water from other sources (by the bottle or truckload). On the surface, it might appear that this is purchasing the same product from a higher-priced source. However, the sectoring scheme of our model, and most multisectoral models, distinguish commodities according to the sector providing them. The sources of supply are very important in disaster impact analysis because of variations in vulnerability across them. For example, bottled and trucked water is less vulnerable to terrorist attack than a below-ground water service utility pipeline network. Applicability is not immediate as in the other options because it takes time to access substitutes. Adaptation can increase somewhat as more substitution options are determined.

5. Back-up supplies. This includes adjustments that incur costs, such as digging of wells, and low cost measures, such as collecting rainfall or using riverine water. The costless alternative sources can be modeled in a manner similar to conservation, and the cost-incurring ones in a manner similar to substitution. The use of water inventories (stored water) is best addressed by the previous response. The response is often implemented in the immediate aftermath of an earthquake through the long run for some of the measures.

This section and the previous one have invoked the assumption of optimizing behavior and implicitly assumed an absence of uncertainty regarding consumer response options and the outcome of their implementation. However, the analysis can be generalized to include features of bounded rationality (for example, satisficing, or even panic, behavior) and uncertainty.

HOUSEHOLD PRODUCTION FUNCTIONS IN THE CONTEXT OF CATASTROPHES

The contribution of the HPF approach to consumer behavior specification in the economic analysis of disaster events is similar to its role in environmental valuation. The HPF introduces the needed structure in valuation by providing economists with a framework for assigning roles to environmental goods in consumption. In other words, it enables the modeler to define relationships among market goods, time and other non-market goods by introducing 'technologies of consumption'. The focus of household behavior analysis becomes that of maximizing household utility from services flowing from those goods they 'produce' for consumption rather than the conventional approach that treats market purchases as direct sources of utility. Thus, household responses to market and physical changes in the economy are subject to intervening household technology processes before affecting welfare. It is this intervening process that allows for resilience, as distinguished from responses to market prices and incomes changes associated with conventional consumer behavior specification. This is especially important under catastrophe scenarios where the breakdowns of market processes bring endogenous household responses to the fore. The potential of the HPF approach for illuminating the economic consequences of catastrophes is faced with difficulty due to fact that household service quantities and shadow prices are unobservable; akin to modeling market production without observations on output quantities and prices. Although certain plausible assumptions allow use of the HPF to proceed in the face of these shortcomings, it means that HPF specifications have to be adapted to specific modeling needs.

Although the Katrina disaster's scale distinguishes it from those that have received the attention of most analysts, there is yet another feature that separates it from other disasters. Hurricane events such as Katrina are characterized by forewarnings that enable the government to issue an evacuation order, as opposed to the sudden occurrence of an earthquake or other events. However, not all residents left the affected areas before the hurricane struck. Although some residents chose to stay in the city, many of the residents caught by the hurricane had no means to leave. This dichotomy in the population response to the evacuation order is an important aspect of the impacts of the Katrina disaster. An HPF specification could help provide some insight into this response by differentiating residents in terms of dependence and access to public or private transportation in the bid to evacuate the city. It would be possible to examine the influence of the government's decision and ability to increase the public transport system capacity for evacuation purposes.

Another crucial impact of the Katrina disaster is the widespread destruction of housing stock. An HPF formulation would need to distinguish the role of this stock in household behavior as contrasted with its usual treatment as flows from the real estate sector of the market economy. For example, the income balance of many households is tied to the value of the housing stock. Therefore, an HPF specification that accounts for the various roles of the housing stock in household behavior would account for its influence on the ability of households to meet other expenses, including matters relating to the ability to obtain credit. Under such a formulation, temporary shelters could be modeled as government-provided alternatives to the lost housing stock. Also, the welfare effects of insurance payments and government support payments that often follow such events can be explicitly captured. The CGE model can then estimate the broader regional economic impacts of these inflows of funds.

Major catastrophes such as Katrina also impress upon us the importance of 'life safety' issues and the need to model them. Safety can also be considered a household 'produced' good, combining purchased goods, time and attributes such as vigilance. In fact, ultimately, we may even consider moving up the hierarchy and listing 'survival' as a household-produced good, which combines utility lifeline services, food, safety and survival characteristics. Resilience plays a prominent role here as well.

Finally, we note that catastrophes pose an especially strong challenge to resilience and may in fact erode it significantly. The high levels of resilience mentioned earlier are based on more limited circumstances, even in the context of disasters. A terrorist attack on a water system leaves factories unscathed and readily able to reschedule lost production once the contaminant has been neutralized, and households can also make up lost production. Most natural diasters, even large ones such as the Northridge earthquake (Tierney, 1997; Rose and Lim, 2002), have also caused limited damage to production units (and to homes) relative to the total population of the relevant region. In the case of Katrina, however, the devastation was widespread. Not only were the majority of lifeline service companies in the New Orleans area out of operation, but businesses and households sustained so much physical damage and social disarray as to render them dysfunctional even after water and power were restored. As Rose (2008) has described in detail for the case of businesses, a catastrophe is likely to erode significantly each type of resilient response. The magnitude, duration and geographic scope of a disaster are similarly likely to influence household resilience as well. Fortunately, these effects can be modeled by including housing stock and other key variables in the household production function.

CONCLUSION

The economic impact of natural and man-made hazards on households is potentially enormous and should no longer be neglected. In this chapter, we have presented an operational approach for estimating these effects by way of a household production function (HPF), which combines purchased goods, non-market goods, time and ingenuity to yield conventional household goods, such as boiled or decontaminated water, as well as less tangible but critical endeavors, such as safety and even survival. This formulation is also able to capture resilience relating to normal behavior and ingenuity in the face of a crisis. The HPF approach provides several valuable features including a formalization of the household decision-making process, a checklist of household needs and attributes, a way of valuing non-market goods and services (including household time), and the contribution of ingenuity, including survival skills.

NOTES

* The authors wish to acknowledge the funding support of CREATE, as well as related funding from the NSF-sponsored Multidisciplinary Center for Earthquake Engineering Research (MCEER).
1. These measures stem from welfare economics, an analytic framework for the valuation of economic outcomes or policies. 'Welfare' in this context refers to general economic well-being and not to specific government transfer programs such as food stamps or unemployment insurance. Note also that greater detail in the household sector facilitates the calculation of the distribution of impacts of terrorist attacks across socio-economic groups in terms of both conventional income and welfare measures.
2. Actually, most of the goods are exchanged through a market in a similar manner to final goods and services, but the 'market' box is omitted in Figure 11.1 to avoid cluttering it. Note also that a significant amount of intermediate goods are not transmitted through a market in the case of vertically integrated firms.
3. Note another simplification of the diagram is the omission of the government sector, which produces many goods and services utilized by both households and businesses. In an increasing number of cases, these goods and services are transacted through markets (for example, water and electricity), but the majority are still provided indirectly through the tax/expenditure system, which does not match payments with service levels provided.

4. This definition subsumes whether a system can 'snap back' at all, that is, the concept of 'stability' as typically used in dynamics. 'Desired state' is a generalization of possible responses, which would include pre-disaster status as a special case, but would at the same time allow for growth and change over time, as well as including obstacles to achieving the desired state.
5. For a discussion of business response to water disruption see Rose and Liao (2005), and for discussion of business response to electricity disruptions see Rose et al. (2007a).

REFERENCES

Barnett, W. (1977), 'Pollak and Wachter on the household production function approach', *Journal of Political Economy*, **85**, 1073–82.
Becker, G.R. (1965), 'A theory of the allocation of time', *Economic Journal*, **75**, 493–517.
Brookshire, D. (2005), 'Economic impacts of terrorist attacks on iconic targets', paper prepared for US Department of Homeland Security, Department of Economics, University of New Mexico.
Bruneau, M., S. Chang, R. Eguchi, G. Lee, T. O'Rourke, A. Reinhorn, M. Shinozuka, K. Tierney, W. Wallace and D. von Winterfeldt (2003), 'A framework to quantitatively assess and enhance seismic resilience of communities', *Earthquake Spectra*, **19**, 733–52.
Chiappori, P. (1997), 'Introducing household production in models of family labor supply', *Journal of Political Economy*, **105**, 191–209.
Cho, S., P. Gordon, J. Moore II, H. Richardson, M. Shinozuka and S. Chang (2001), 'Integrating transportation network and regional economic models to estimate the costs of a large urban earthquake', *Journal of Regional Science*, **41**, 39–65.
Comfort, L. (1999), *Shared Risk: Complex Seismic Response*, New York: Pergamon.
Copeland, C. (2006), 'Hurricane-damaged drinking water and waste water facilities: impacts, needs, and response', CRS Report for Congress, Washington, DC: Congressional Research Service, 24 May.
Gordon, P., J. Moore II, H. Richardson and Q. Pan (2005), 'The economic impact of a terrorist attack on the twin ports of Los Angeles–Long Beach', in H. Richardson, P. Gordon and J. Moore II (eds), *The Economic Impacts of Terrorist Attacks*, Cheltenham, UK and Northampton, MA, USA: Edward Elgar, pp. 262–86.
Heinz Center for Science, Economics and the Environment (2000), *The Hidden Costs of Coastal Hazards: Implications for Risk Assessment and Mitigation*, Washington, DC: Island Press.
Holling, C. (1973), 'Resilience and stability of ecological systems', *Annual Review of Ecology and Systematics*, **4**, 1–23.
Hori, H. (1975), 'Revealed preference for public goods', *American Economic Review*, **65**, 978–91.
Kerkhofs, M. and P. Kooreman (2003), 'Identification and estimation of a class of household production models', *Journal of Applied Econometrics*, **18**, 337–69.
Mendelsohn, R. (1984), 'Estimating the structural equations of implicit markets and household production functions', *Review of Economics and Statistics*, **66**, 673–77.
Muth, R. (1966), 'Household production and consumer demand functions', *Econometrica*, **34**, 699–708.
Oladosu, G. (2000), 'A non-market computable general equilibrium model for economic analysis and climate change in the Susquehanna River Basin', PhD thesis,

Department of Energy, Environmental, and Mineral Economics, Pennsylvania State University, University Park, PA.

Oladosu, G. (2006), 'Construction of household production functions for use in CGE modeling: an application to terrorist events', Environmental Sciences Division, Oak Ridge National Laboratory, Oak Ridge, TN.

Pollak, R. and T. Wales (1992), *Demand System Specification and Estimation*, New York: Oxford University Press.

Pollak, R. and M. Wachter (1975), 'The relevance of the household production function and its implication for the allocation of time', *Journal of Political Economy*, **83**, 255–77.

Rose, A. (2004), 'Defining and measuring economic resilience to disasters', *Disaster Prevention and Management*, **13**, 307–14.

Rose, A. (2005a), 'Analyzing terrorist threats to the economy: a computable general equilibrium approach', in H. Richardson, P. Gordon and J. Moore, II (eds), *Economic Impacts of Terrorist Attacks*, Cheltenham, UK and Northampton, MA, USA: Edward Elgar, pp. 196–217.

Rose, A. (2005b), 'A typology of economic disruptions', Department of Geography, Penn State University, University Park, PA.

Rose, A. (2008), 'Macroeconomic impacts of catastrophic events: the influence of resilience', in J. Quigley and L. Rosenthal (eds), *Risking House and Home: Disasters, Cities, Public Policy*, Berkeley, CA: Berkeley Public Policy Press.

Rose, A.J. Benavides, S. Chang, P. Szczesniak and D. Lim (1997), 'The regional economic impacts of an earthquake: direct and indirect effects of electricity lifeline disruptions', *Journal of Regional Science*, **37**, 437–58.

Rose, A. and S. Liao (2005), 'Modeling resilience to disasters: computable general equilibrium analysis of a water service disruption', *Journal of Regional Science*, **45** (1), 75–112.

Rose, A. and D. Lim (2002), 'Business interruption losses from natural hazards: conceptual and methodology issues in the case of the Northridge earthquake', *Environmental Hazards: Human and Social Dimensions*, **4**, 1–14.

Rose, A., G. Oladosu and S. Liao (2007a), 'Business interruption impacts of a terrorist attack on the electric power system of Los Angeles: customer resilience to a total blackout', *Risk Analysis*, **27** (3), 513–31.

Rose, A., G. Oladosu and S. Liao (2007b), 'Regional economic impacts of a terrorist attack on the water system of Los Angeles: a computable general disequilibrium analysis', in P. Gordon, J. Moore and H. Richardson (eds), *Economic Costs and Consequences of a Terrorist Attack*, Cheltenham, UK and Northampton, MA, USA: Edward Elgar.

Schlenger, W., G. Norton, D. Walker, A. Goldberg and C. Lewis (2006), 'Estimating loss of life from hurricane-related flooding in the greater New Orleans area: health effects of Hurricane Katrina', unpublished paper.

Shinozuka, M., S. Chang and A. Rose (1998), 'Infrastructure life cycle cost analysis: direct and indirect user costs of natural hazards', in M. Shinozuka and A. Rose (eds), *Proceedings of the US/Japan Joint Seminar on Civil Infrastructure Systems Research*, Buffalo, NY: MCEER.

Tierney, K. (1997), 'Impacts of recent disasters on businesses: the 1993 Midwest floods and the 1994 Northridge earthquake', in B. Jones (ed.), *Economic Consequences of Earthquakes: Preparing for the Unexpected*, Buffalo, NY: National Center for Earthquake Engineering Research.

12. Adjusting to natural disasters

**V. Kerry Smith, Jared C. Carbone,
Jaren C. Pope, Daniel G. Hallstrom and
Michael E. Darden**

INTRODUCTION

Natural disasters force adjustment. The Indian Ocean tsunami in late 2004 and Katrina's devastation in New Orleans and throughout the Gulf Coast in August 2005 renewed our collective awareness of disasters. Natural disasters can also provide opportunities for quasi-random experiments to learn how households adjust to these events. For example, Kahn's recent (2005) analysis of the death toll from natural disasters suggests that the level and distribution of income, along with presence of 'higher-quality' institutions, can influence the severity of the outcomes from disasters. Nonetheless, there are limits to our ability to use a detailed specification for a reduced form model to parse the factors in many different countries that contribute to social and economic outcomes. Ideally, one would like to be able to hold political institutions and the cultural context constant.[1] However, this strategy forfeits the cross-country variation providing detailed information of the spatial differences in disasters. As a rule, detailed spatial information of this type is unavailable for a single location. Our analysis is unique in that we have a spatially delineated record of *ex post* damage of a large-scale disaster as well as maps designating risk zones that proxy for *ex ante* hazard information for our study area.

Our analysis is also related to the literature in economic geography spawned by analyses summarized in Krugman (1998). Most empirical tests of this framework focus on the importance of increasing returns for the degree of spatial differentiation. For example Davis and Weinstein (2002) consider Allied bombing of Japanese cities in World War II as a shock to relative city sizes and find that the location of densely settled areas is preserved even in the presence of significant temporary shocks of a large scale. Comparable conclusions were drawn by Miguel and Roland (2006) for the bombing's effect on local poverty, consumption levels and population density in Vietnam 25 years after the Vietnam War. However in this case there was significant effort at reconstruction of bombed facilities. These analyses are primarily relevant for the size and density of population, not the mix by demographic group or the distribution of income and housing values.

Hurricane Andrew made landfall in August 1992 and was the largest US natural disaster on record prior to Katrina. With the 1990 and 2000 Censuses, we consider how people adjusted to the damage caused by the storm.[2] This can happen in a number of ways: people can move out of harm's way; they can self-protect, building structures less vulnerable to damage; or they can insure. In the first part of our results, we provide a description of how Andrew reshaped Dade County, Florida after the hurricane made landfall and destroyed a large portion of the private housing and commercial facilities. The second part of the analysis uses Census data on the value of homes in areas likely to be perceived as subject to higher risk after the disaster. This component focuses on the market consequences of these adjustments. Both parts provide insight for designing policies that facilitate people's ability to return to their everyday activities after disasters.

Our empirical models exploit a unique *ex post* evaluation of Andrew's damage conducted by the National Oceanic and Atmospheric Administration (NOAA). The findings were published in the *Miami Herald* on 20 December 1992 (referred to later as the NOAA/*Miami Herald* data) (NOAA, 1992). They include information on 420 subdivisions and condominium developments in the area affected by Andrew. We also acquired the Federal Emergency Management Agency (FEMA) flood maps providing a record of the differing nature of the *ex ante* flood risk of damage throughout the country. By matching these two data sources with demographic and economic measures for Dade Country block groups, a spatial picture develops of both adjustments that took place and the context for those changes.

Our findings confirm some prior beliefs about this hurricane and overturn others. In contrast to the popular views of the storm's impact, white, middle-income households experienced more significant damage than poor minority households. Financial capacity, as reflected by home ownership and education, are key factors in who adjusted to the damage. In the eight years after Andrew the population in areas with 50 percent or more of the homes damaged so seriously as to be rated uninhabitable grew faster than areas with less damage. There does not appear to be a significant pattern of adjustment for white and black homeowners in relation to the damaged areas. White renters moved away from damaged areas. Hispanic households, both owners and renters, moved into the areas with hurricane damage. Lower-income households tended to move into damaged areas while middle income moved out. In general, the storm's damage did not affect higher-income households. In 2000, households with annual incomes over $150 000 were the only group likely to be attracted to areas with a comparable 'type' of household – that is, to areas where households with the same income level lived in 1990. Thus, the analysis highlights the potential importance of household heterogeneity for measuring the effects of spatially delineated environmental impacts. Indeed we suggest that studies based on summary measures, such as the median income, miss much of the story.[3]

The second half of the analysis – the evaluation of the economic consequences of these adjustments – evaluates the changes in the distributions of rents and homeowners' beliefs about their homes' values between the 1990 and 2000 Censuses in response to average damage and the fraction of a block group in a FEMA risk zone. The coefficients for the areas with differing risks of coastal flooding in models using changes in the median measures for housing values indicate slower appreciation in both the high- and the medium-risk areas (based on the FEMA flood ratings). The difference in the coefficients for the areas is not statistically significant. However, when we weight each coefficient by the average land area in each Census block group in the risk category, the differences are significant and the relative magnitudes of the effects are consistent with the ordering of the risks. Using the average share of the block groups in the higher risk, this information would imply about a 3 percent reduction in the median value, while for the medium-risk areas it would be about 1 percent due to the information we hypothesize was acquired from the storm.

The next section develops the hypotheses motivating our expectations for differences in households' adjustments. The third section describes the spatially delineated data required to undertake the analysis. Our results are developed in the following section in two parts. First we describe the changing features of neighborhoods based on the 1990 and 2000 Censuses for the county. After that we discuss the changes in the distribution of housing values and rents with the location of the block groups in relation to FEMA flood risk zones and the NOAA measures for Andrew's damage. The last section discusses the implications of the analysis for what we might expect for the pattern of adjustment in the Gulf Coast area after Katrina.

NON-MARKET AND MARKET RESPONSES

Background

When a large share of the private and public capital supporting daily activities is significantly damaged, some private responses are inevitable. Kahn (2005) argues that greater political accountability will induce democracies to take proactive actions and adapt to hazards in a way that reduces their overall impact. We know very little about who is adjusting and what factors induce those who do to act. Most of the available economic models of household adjustment to exogenous changes in community attributes are intended to describe responses to relatively small changes in features of a home or a neighborhood. In the empirical tests of these models the attributes of interest are assumed to be conveyed to homeowners through their locational choices. Households are assumed to be heterogeneous with different preferences for location-specific amenities. As a rule, they assume there is one or more endogenous (to the adjustment process) attributes of neighborhoods that can reinforce or retard responses to an exogenous change in the location-specific attribute. For example, in the externality/filtering models (Coulson and Bond, 1990) average neighborhood income is hypothesized to be a factor that influences household preferences for a neighborhood. It also changes as people alter their choices for neighborhoods. As a result, changes in mean income can enhance or reduce the effects of an exogenous change in a neighborhood attribute. The overall effect this process has on composition of an area depends on the size and direction of the effect of neighborhood mean income on the marginal willingness to pay for the attribute that changes.

A comparable set of influences can be found in the sorting models that would have consistent predictions for large changes in neighborhood attributes and indeterminant implications for small (see Banzhaf and Walsh, 2004). The externality/filtering and sorting models rely on a common formal structure. The first identifies two types of households who must select among locations with continuous variation in an exogenous attribute. Sorting models assume a finite set of communities and continuous variation in household tastes. Banzhaf and Walsh illustrate their analysis with two communities varying in an exogenous attribute and describe sorting among communities. In both structures an equilibrium is defined. With filtering the equilibrium definition stems from the law of one price, whereas with sorting it is boundary indifference. Comparative statics with each relationship, given the constraints linking household heterogeneity to endogenous outcomes, yields the implications for how heterogeneity in preferences or constraints affects the impact of a change in the exogenous attribute. Both models require some version of the single crossing condition and a large change to derive unambiguous hypotheses about outcomes. This requirement for a large change is a key advantage for analysis of outcomes after natural disasters.

The filtering and sorting frameworks have two implications for our research. First, the larger the damage in a neighborhood, the greater the prospects for a change in measures of its demographic and economic composition. Such changes offer the opportunity, given that individuals have the resources to pay for adjustment, to observe whether the exogenous change offsets any endogenous retarding (or enhancing) effects of the changes in the existing composition of a neighborhood. Second, the models imply that uncovering the effects of damage adjustment requires comparing changes in the distributions of the household types with the 1990 and 2000 Censuses rather than changes in measures of the central tendencies for these distributions.

Models

Our strategy for tracking who adjusts uses a simple regression format. We estimate how $(y_{j(t+10)} - y_{jt})$ varies with the average proportion of homes that are judged uninhabitable. y_{jt} is the proportion of households (or individuals depending on the measure being summarized) in category j for $t =$ 1990. For example this could be the proportion in a racial group or it could be the proportion born in Florida. This relationship is estimated with a variety of control variables, including the baseline (that is, 1990) proportion of households (or individuals) in each group, the location of block groups in relation to FEMA flood zones, and the potential effects of a neighborhood bordering Homestead Air Force Base.[4]

An expanded specification for the basic model is given in equation (12.1) with d_j designating the *Miami Herald* damage measure and z_{jk} a set of variables that correspond to the different controls investigated as part of evaluating the robustness of our conclusions. To test the logic of Brock and Durlauf's (2001) social interactions models, we consider some models with z_j corresponding to the y_{jt}. The argument is simply that demographic groups seek to stay within areas that share common interests and networks. For other models, z_{jk} corresponds to measures of the extent of a block group in areas with higher risks of coastal flooding as measured by the FEMA flood maps.

$$\left(y_{j(t+10)} - y_{jt}\right) = \alpha_0 + \alpha_1 d_j + \sum_k \tau_k z_{jk} + \varepsilon_j \qquad (12.1)$$

where:

ε_j is a random error assumed to be classically well behaved
α_0, α_1, and τ_k are parameters to be estimated

The cell definitions of some of the economic variables, such as the distributions of incomes, rents and housing values, changed between the two Censuses. These changes in the 2000 Census expand the resolution in the middle categories and change the upper censoring point. We redefined the 2000 categories for rents, housing values and income so they matched the 1990 categories.

For property values and rents these models generalize the logic proposed by Chay and Greenstone (2005) to use quasi-random experiments to estimate the incremental value of changes in site specific amenities. That is, a hedonic model's ability to recover an estimate on the incremental willingness to pay for amenities conveyed through residential location relies on sorting behavior. People select the best locations they can afford. Nonetheless, there may be unobserved differences in the households selecting locations with low amenity levels in comparison to those with high levels. Use of an exogenous instrument and a difference-in-difference framework allows the effect of interest and the influence of unobserved heterogeneity to be distinguished.

Applications of this logic for environmental effects have generally relied on county (or Census tract)-level mean or median housing values across Censuses and we report the results for these summary measures as well. However, we add to them another approach – examining the changes in the distributions of housing values and rents. To the extent that there is a change in the composition of the housing available as a result of the amenity differences, the 'average' may not be distinguishing a marginal value for the change in the amenity. By using changes in the distributions of housing values or rents, we have greater control over the 'types' of housing through the value and rent brackets. As a result, it is possible to consider how amenity changes influence the composition of housing.

DATA

In December 1992 the *Miami Herald* published a special report analyzing the factors responsible for areas with significant damage that were far from the storm's strongest winds. As part of the report, the newspaper included the full documentation for the NOAA damage assessment by local housing subdivisions. Using the map included with the *Miami Herald*'s feature article it was possible to align the roadways with an Arcview map of the primary roads within the county. A set of 306 grids was defined to match the subdivision records to Census block groups. Each block group was assigned the average damage measure for the subdivisions falling within its boundary (area weighted if a subdivision crossed Census boundaries).

These damage estimates are proxies for the extent to which neighborhoods offer opportunities for nearly complete replacement of residential structures. When 100 percent of the homes in a neighborhood are judged to be uninhabitable, then it seems clear the damage measure offers a clear-cut index of the opportunities to transform the composition of the area. The impact of smaller amounts of damage on changes in the composition of a block group depends on several factors. Partial damage may well signal the quality of the remaining housing stock. In fact, an important motivation for the *Miami Herald*'s special report was the heterogeneity in damage by subdivision. It was not completely consistent with the hurricane's areas with highest wind levels from the hurricane. To develop this comparison, we obtained Wakimoto and Black's (1994) maps describing the wind patterns for Andrew. A cross-tabular analysis of the *Miami Herald* damage survey data with an approximate wind-based qualitative variable for the damage suggests higher wind areas were more likely to experience damage. Nonetheless, as the *Miami Herald* story 'Less winds, lots of damage', 20 December 1992 (NOAA, 1992), documents there are a number of exceptions. Moreover, the *Miami Herald* feature also identified problems in the county's building inspections, noting that: 'Unsupervised and understaffed, with civil service rules that give them job protection, Dade's building inspectors were no match for the development of the 1980s.'[5]

Actual implementation of a spatial analysis requires a number of judgments. For example, between the 1990 and 2000 Censuses, the definition of block groups for the county changed, expanding from 1048 in 1990 to 1222 in 2000. This reflects the increase in population in the county and the need to realign Census summaries to the population growth. To avoid mixing the potential for endogeneity in the neighborhood definition with the event being studied (that is, Andrew's damage) we map the 2000 records into the 1990 definition of block groups.

We construct area weighted averages of Census statistics from the 2000 block groups so that each record can be matched to its 1990 counterpart. Table 12.1 reports an overall summary of the demographic and economic patterns in 1990 and 2000. The average proportion of each demographic and economic category across block groups is reported for three samples. The last two columns labeled 'overall' provide these average proportions for all the block groups between 1990 and 2000. The first two sets of columns decompose this set into block groups experiencing 50 percent or greater of their homes as uninhabitable based on the NOAA/*Miami Herald* survey. The number of block groups in this category is 27. The second group includes those with less than 50 percent uninhabitable. The number of observations in this category ranges from 968 to 997 depending on the variable selected.

Comparing the attributes of the populations in the damaged areas in 1990 to summary measures of these attributes for the County as a whole in

Table 12.1   The composition of Dade County by damage class from Andrew, 1990 and 2000 Censuses

	Greater than 50% uninhabitable		Less than 50% uninhabitable		Overall summary	
	1990	2000	1990	2000	1990	2000
I. Demographic composition						
Owner occupied						
White	0.780	0.705	0.716	0.663	0.718	0.664
Black	0.140	0.175	0.236	0.258	0.234	0.256
Hispanic	0.200	0.439	0.444	0.520	0.437	0.518
Renters						
White	0.714	0.549	0.638	0.605	0.640	0.604
Black	0.201	0.308	0.291	0.290	0.289	0.291
Hispanic	0.261	0.428	0.507	0.541	0.501	0.538
II. Income distribution						
< 15 000	0.189	0.175	0.317	0.247	0.313	0.245
15 000–25 000	0.138	0.167	0.178	0.148	0.177	0.149
25 000–40 000	0.229	0.198	0.196	0.177	0.197	0.178
40 000–60 000	0.284	0.174	0.151	0.159	0.155	0.159
60 000–150 000	0.149	0.256	0.131	0.216	0.132	0.217
> 150 000	0.011	0.031	0.026	0.052	0.026	0.052

1990 suggests the hurricane's damage was not disproportionately experienced by minority or poor households. In 1990 block groups with 50 percent or over damage were largely white (both owner and rental households) in the income range from $25 000 to $60 000. When we consider the proportions in 2000, white households appear to have moved out (both owners[6] and renters). Hispanic households moved in. These changes partially reflect the overall growth in the share of Hispanic households in the county. To the extent that they are moving from outside the US, it may also reflect differences in the information they have about the risks in these areas.

Based on these averages it appears that middle-income households moved out and the lower-($15 000–$25 000) and high-income groups moved in. The results would change somewhat if we modify the threshold used to isolate the high-damage block groups. Our separation at 50 percent leads to a relatively small sample of block groups that underlies the means used to characterize who is adjusting to extensive damage. Below, we use regression models to evaluate how the differences in damage at block groups affect changes in their composition.

RESULTS

Who Adjusts

Our analysis of the changes in the composition of the Census blocks in Dade County between 1990 and 2000 considers three types of models and reports for some of these our more comprehensive evaluation of the effects of the area definition (for example 1990 versus 2000 block groups). The first set of models evaluate whether the proportionate change in a demographic or economic variable describing population changes are related to the NOAA/*Miami Herald* damage measure assigned to each block group. These analyses include such dependent variables as the counts of white, black and Hispanic homeowners or the households in the $40 000 to $60 000 income bracket, and so forth. As noted, these models are estimated with two samples. The first uses the 1990 block definitions and the full sample of block groups. Different area weights are used in the counts and are continuous variables.[7]

The second sample is intended to evaluate the effect of the area weights used to reconstruct the 1990 equivalents. For these analyses, we use a sample with only the block groups that did not change between 1990 and 2000. The second group of models evaluates whether the relationships between the proportionate changes in the variables measuring demographic attributes and the NOAA/*Miami Herald* damage measure depend on the initial (that is, in 1990) fraction of each group in each 1990 block group. This strategy offers a simple gauge of whether the social interactions logic influences the relationship we observe between damage and the change in groups. Finally, the third set of analyses considers whether the FEMA flood zones influence the locational choices of different groups.

Table 12.2 reports the simple model, considering whether the fraction of households reporting that they stayed in the same house was influenced by the NOAA/*Miami Herald* damage measure assigned to each block group. Damage did not affect the propensity to leave one's house or county. It does appear to influence the relocation patterns of those households born outside Florida. These groups avoid areas with damage. Native Floridians are then a disproportionately higher share of the population. None of these results is affected by which sample was used for the tests.

Table 12.3 reports the simple models for demographic variables, income, rents and housing values. Each entry in the table corresponds to a different model where the dependent variable is the proportionate change between the 1990 and 2000 Censuses and the independent variable is the NOAA/*Miami Herald* damage measure (d_j in equation 12.1) or this measure along with the 1990 proportion of the relevant group in each block group (WI for 'with initial conditions'). We do not report estimates for the parameters

Table 12.2   Types of adjustments in response to Andrew's damage

Model	Same Block group	Area weighted 2000–1990
A. *STAYING PUT*		
Proportion – same house	−0.04	0.00
	(−0.88)	(0.03)
Proportion – same county	0.04	0.04
	(1.00)	(1.37)
B. *CHANGES IN COMPOSITION BASED ON BIRTH AREA*		
Midwest	−0.03	−0.04
	(−2.54)	(−4.75)
Northeast	−0.02	−0.03
	(−1.16)	(−1.61)
South	−0.10	−0.08
	(−4.97)	(−4.37)
West	−0.01	−0.00
	(−0.79)	(−0.30)
Florida	0.06	0.06
	(1.84)	(2.36)

Note: The numbers in parentheses are t-ratios for the null hypothesis of no association.

associated with this baseline proportion. The table entries indicate its sign (N or P) and significance (S or I) as a gauge for the robustness of the estimates for the damage measure.

White renters appear to avoid damaged areas. It appears that black households with home equity adopt the same adjustments in qualitative terms as the white owners – moving away from damaged areas – but neither model has these negative effects statistically significant. Black renters and Hispanic households, both owners and renters, increase in the damaged areas. While some of the Hispanic increase reflects an overall increase in this demographic group, as suggested in the average proportionate growth measures by demographic groups for the county as a whole (in Table 12.1), there is also a disproportionate growth in the damaged areas. Considering the results for groups based on the various educational levels achieved, the proportions with less than high school education along with those who have graduate degrees are consistently significant and negatively related to the damage measure.[8]

Use of the proportionate changes in the groups in the income cells allows more direct consideration of the heterogeneity arguments associated with the tipping/sorting models used to describe how the composition of a community changes in response to an exogenous shock. The proportion of

Natural disaster analysis after Hurricane Katrina

Table 12.3 Demographic and economic adjustments to Andrew's damage

Model	Same block group		Area weighted 2000–1990	
	S	WI*	S	WI[b]
A. DEMOGRAPHIC				
1. Owner Occupied				
Proportion – White	−0.09	−0.01 N	−0.41	−0.04 N
	(−0.21)	(−0.25) S	(−1.16)	(−1.15) S
Proportion – Black	−0.05	−0.47 N	0.01	0.01 N
	(−1.42)	(−1.42) S	(0.34)	(0.20) S
Proportion – Hispanic	0.24	0.18 N	0.23	0.16 N
	(6.19)	(4.76) S	(6.72)	(4.94) S
2. Renters				
Proportion – White	−0.95	−0.11 N	−0.13	−0.12 N
	(−1.79)	(−2.14) S	(−2.91)	(−3.07) S
Proportion – Black	0.12	0.12 N	0.13	0.12 N
	(2.86)	(2.98) S	(3.64)	(3.66) S
Proportion – Hispanic	0.18	0.07 N	0.22	0.12 N
	(2.93)	(1.29) S	(4.55)	(2.62) S
B. EDUCATION				
Proportion less than H.S.	−0.21	−0.21 P	−0.21	−0.20 P
	(−2.99)	(−2.98) S	(−3.63)	(−3.57) I
Proportion with H.S.	−0.05	−0.06 N	−0.04	−0.03 N
	(−1.49)	(−1.91) S	(−1.73)	(−1.40) S
Proportion with some college	−0.00	−0.03 N	−0.01	−0.02 N
	(−0.18)	(−1.30) S	(−0.63)	(−1.17) S
Proportion with college	−0.04	−0.04 P	−0.03	−0.03 P
	(−2.00)	(−1.87) I	(−1.72)	(−1.57) S
Proportion with graduate school	−0.06	−0.07 N	−0.05	−0.06 N
	(−2.10)	(−2.30) S	(−2.44)	(−2.66) S
C. INCOME				
Proportion Income < 15K	0.09	0.09 N	0.11	0.10 N
	(2.63)	(2.54) S	(3.47)	(3.24) S
Proportion 15K < Income < 25K	0.06	0.06 N	0.05	0.05 N
	(1.85)	(1.99) S	(1.91)	(1.90) S
Proportion 25K < Income < 40K	0.02	0.04 N	−0.02	−0.02 N
	(0.64)	(1.04) S	(−0.66)	(−0.57) S
Proportion 40K < Income < 60K	−0.16	−0.14 N	−0.17	−0.16 N
	(−5.39)	(−4.48) S	(−6.79)	(−6.51) S
Proportion 60K < Income < 150K	−0.04	−0.03 N	0.03	0.03 N
	(−1.03)	(−0.81) S	(1.02)	(1.10) S

Table 12.3 (continued)

Model	Same block group		Area weighted 2000–1990	
	S	WI*	S	WI[b]
Proportion Income > 150K	0.03	0.03 P	0.01	0.01 P
	(1.38)	(1.56) S	(0.52)	(0.77) S
D. *RENTS*				
Proportion Rent < 250	−0.05	−0.03 N	−0.06	−0.04 N
	(−1.10)	(−0.81) S	(−1.61)	(−1.31) S
Proportion 250<Rent < 500	0.26	0.23 N	0.31	0.26 N
	(3.29)	(3.06) S	(4.70)	(4.18) S
Proportion 500<Rent < 750	−0.20	−0.21 N	−0.22	−0.23 N
	(−2.19)	(−2.20) S	(−2.76)	(−2.93) S
Proportion 750<Rent < 1000	−0.02	−0.02 N	−0.12	−0.12 N
	(−0.23)	(−0.20) I	(−1.87)	(−1.88) I
Proportion Rent>1000	0.01	−0.00 N	0.09	0.09 P
	(0.13)	(−0.03) S	(1.55)	(1.59) I
E. *HOUSING VALUES (HV)*				
Proportion HV<40K	−0.04	0.06 N	−0.03	0.04 N
	(−0.63)	(1.16) S	(−0.59)	(0.83) S
Proportion 40K < HV < 100K	0.36	0.47 N	0.19	0.27 N
	(2.96)	(4.30) S	(2.02)	(3.01) S
Proportion 100K < HV < 250K	−0.35	−0.33 N	−0.16	−0.16 N
	(−3.16)	(−3.13) S	(−1.95)	(−1.96) S
Proportion 250K < HV < 400K	0.02	0.02 N	0.00	0.00 N
	(0.46)	(0.47) I	(0.06)	(0.07) I
Proportion 400K < HV < 500K	0.01	0.01 N	0.00	0.00 N
	(0.75)	(0.73) I	(0.23)	(0.22) I
Proportion HV>500K	−0.00	−0.00 P	−0.00	0.00 P
	(−0.07)	(−0.01) S	(−0.14)	(0.02) S

Notes: The numbers in parentheses are t-ratios for the null hypothesis of no association.
* This column corresponds to models that include the damage measure for Andrew along with the initial count of the group being modeled in each block group in 1990. The letters refer to the sign and significance of a term included to reflect the count of households in the relevant group in 1990.
N = negative, P = positive, S = significant, I = insignificant.

households in the lowest two income categories (less than $15 000, and $15 000 to $25 000) grows while the middle-income group ($40 000 to $60 000) declines. Upper-income groups do not significantly change in response to the damage areas. This pattern is broadly consistent with the expectations of sorting models. In a Tiebout (1956) model households adjust to local public goods (and bads) based on their ability to pay. The middle-income group may have the ability to pay for adjustment. Moving to avoid risk is the way they appear to adjust. Lower-income groups may be taking advantage of the lower rents in these areas. They do not have the ability to pay for moving out to another lower-risk area as an adjustment. The damage and reconstruction creates an opportunity when the replacement of residential structures is with lower-cost units.

Higher-income households have the ability to self-protect and to insure. As a result, it seems reasonable to expect a wider array of adjustment possibilities. Moving out of an area may be the last alternative for this group. Thus, there is a reasonable explanation for a lack of any changes with this group. The high coastal risk areas also correspond to areas with high coastal amenities. High-income households in these zones may already have self-protected. When we used the median income, the difference in the log of the median income in the two Censuses is negatively related to damage, but not significant at the 10 percent level (p-value is 0.11). The results in Table 12.3 help to explain why.

Rents and housing values adapt to support the changing composition of households in the damaged block groups. The proportion of lower rent units increased in damaged areas and the higher rent decreased. The same effects can be traced in the signs and significance of the owner-reported housing values. The proportion in the range $40 000 to $100 000 increased with damage while those in the $100 000 to $250 000 decreased. There was no change in the proportions in the higher-valued categories with respect to damage.

Table 12.4 considers whether the adjustments are affected by the ability to avoid risky areas. That is, controlling for the average NOAA/*Miami Herald* damage in a block group we consider whether the fraction in the block group in different FEMA risk categories influenced the proportionate changes in each demographic group and income category. These estimates are based on the full sample of block groups. AE is classified the highest-risk category for coastal flooding, AH next highest, and X500 minimal risk. Homestead is a dummy variable to indicate whether the block group bordered the Homestead Air Force Base (= 1 and 0 otherwise). This facility was closed after being completely destroyed by the hurricane. We might expect two influences with this variable. The first is associated with initial land uses around the base and the second with the scale of the effects of damage along with the base closing to depress land values and reduce economic activity.

White renters tend to avoid block groups with the highest risk (AE). Hispanic owners and renters and black renters seem to increase disproportionately in block groups with the highest risk. The results for income

Table 12.4   *Adjustment and risk information*

	NOAA / Miami Herald Damage	FEMA Flood Zones			Homestead Air Force Base
		AE	AH	X500	
A. DEMOGRAPHIC					
1. Owner Occupied					
Proportion – White	−0.28	−0.01	0.01	0.05	−0.18
	(−0.78)	(−0.86)	(0.50)	(2.37)	(−1.76)
Proportion – Black	−0.00	0.01	−0.01	−0.03	0.18
	(−0.02)	(1.10)	(−0.65)	(−1.94)	(2.02)
Proportion – Hispanic	0.22	0.04	0.06	0.02	0.01
	(6.29)	(3.29)	(2.65)	(1.02)	(0.08)
2. Renter					
Proportion – White	−0.12	−0.03	0.02	0.02	−0.04
	(−2.81)	(−2.00)	(0.77)	(0.93)	(−0.43)
Proportion – Black	0.13	0.02	0.02	−0.00	−0.04
	(3.47)	(2.09)	(0.88)	(−0.14)	(−0.47)
Proportion – Hispanic	0.20	0.03	0.04	0.01	0.23
	(4.01)	(2.20)	(1.27)	(0.50)	(2.12)
3. Income					
Proportion Income <15K	0.10	−0.02	0.06	0.06	−0.03
	(3.39)	(−1.58)	(2.88)	(3.37)	(−0.34)
Proportion 15K< Income<25K	0.05	0.00	−0.03	−0.03	−0.01
	(1.95)	(0.32)	(−1.69)	(−2.19)	(−0.16)
Proportion 25K< Income<40K	−0.02	0.01	−0.02	−0.01	−0.01
	(−0.58)	(0.91)	(−1.01)	(−0.82)	(−0.15)
Proportion 40K< Income<60K	−0.17	−0.00	−0.04	−0.02	0.04
	(−6.52)	(−0.46)	(−2.37)	(−1.32)	(0.55)
Proportion 60K< Income<150K	0.02	−0.00	0.03	0.00	0.05
	(1.71)	(−0.26)	(1.84)	(0.29)	(0.60)
Proportion Income> 150K	0.01	0.01	−0.00	0.00	−0.03
	(0.64)	(2.21)	(−0.55)	(0.03)	(−0.78)

Note: The numbers in parentheses are t-ratios for the null hypothesis of no association.

groups are not as clear-cut as they are for the demographic categories. Low-income groups decrease in the block groups with the largest fraction of their area in the high-risk FEMA category (AE) and increase in the AH category. The proportion of middle-income groups is for the most part unrelated to risk measures. There is some evidence of a shift toward risky areas with income in that the coefficient for AH shifts from negatively related to the low risk to positive, with increases in income from the $40 000 to $60 000 class to the $60 000 to $150 000 group. The most intriguing of the estimates is associated with the over-$150 000 group increasing with the area of the block group in the high-risk category. This seemingly counter-intuitive result likely reflects the higher amenity levels associated with these same locations and the ability of this group to self-protect and insure against the risks posed in these areas.

Overall, these results confirm the importance of accounting for house-holds' heterogeneity in understanding their adjustments to disasters. As expected, ability to pay appears to be important to understanding why we observe differences in demographic groups' responses to damage caused by natural disasters. Ethnic attachment to neighborhoods, as hypothesized in social interaction models and as proxied in our analysis by the 1990 pro-portion of each demographic group, does not overturn the results. Hispanic owners respond to damage while white and black homeowners do not. For Hispanic groups, both owners and renters are more likely to move into damaged areas. In their case treating home ownership as a proxy for ability to pay would not allow us to reconcile these findings with what was esti-mated for other groups.

There appears to be an especially interesting story in the changes in the income distributions. Lower-income groups increase in damaged areas and the proportion of middle-income groups decreases, suggesting there is adjustment to both damage and potentially the perception of increased risk. The lower-income groups may be taking advantage of lower rents. Thus, this finding contrasts with Breen and Gupta's (1997) results suggest-ing little change in demographics in response decisions to use new areas for facilities with increased environmental risk.[9]

Our analysis suggests higher income groups do not adjust to the damage caused by disasters and, if anything, tend to move to coastal locations with higher risks of flooding damage. Had we used the change in the median income between the two Censuses, our conclusion would have been much different. It would have suggested the size of the area in high-risk zones was negative and significant influence. Equation (12.2) provides these estimates (with t ratios in parentheses).[10]

$$\ln(\tilde{m}_{t+10}) - \ln(\tilde{m}_t) = 0.298 - 0.19 \ NOAA \ Damage$$
$$(13.14) \ (-1.60) \quad\quad\quad (12.2)$$
$$-0.11 = AE - 0.22 \ AH - 0.10 \ X500$$
$$(-2.95) \quad\quad (-0.29) \quad\quad (-1.58)$$

where:
$n = 1022$
$R^2 = 0.01$

The significant negative coefficient for the proportion of the block group in the AE zone would cause one to overlook the differences in these FEMA zones roles for different income groups. For the highest-income groups, the amenity effect dominates that of risk and the proportion of these households increased in areas with higher risk.

A Quasi-Experimental Analysis of Market Responses to Damage and Risk information

Substantial changes in the housing stock or in spatially delineated amenities or disamenities should be signaled through markets. These effects are, after all, the economist's stock in trade. They are the basis for the quasi-experiments referred to earlier. In some respects our research to this point provides motivation for an analysis of housing prices and rents. Our estimates suggest the event – Hurricane Andrew – did cause adjustment. Households, whether classified by demographics or income, responded. Now we ask whether Census measures of housing values and rents provide consistent signals of these adjustments.

Tables 12.5 and 12.6 report our estimates. In Table 12.5, we estimate the change in the log of the median housing values for 2000 compared to 1990 and the change in the median rents. In each case we evaluate models with controls for changes in the composition of the 'average' units along with the damage measure and proportion of the block group in each FEMA flood risk zone. Because the model considers the proportionate change (that is, differences in the log of the median values) in housing values and rents, we hypothesize that these variables capture the change in perceptions of the risk (for the FEMA variables) as a result of Andrew and the net effect of changed quality perceptions and damage related changes to the composition of housing markets for the *Miami Herald* variable.

The estimated coefficients for the damage measure are insignificant in all models – implying that the market effects of the compositional changes we observed from demographic measures are not capable of being recovered with these regressions using proportionate changes in medians. The FEMA flood risk measures are consistent with our hypothesis that they reflect the effects of changed risk perceptions for these areas due to Andrew. The proportions of the Census block group in both the highest-risk (AE) and the next highest-risk (AH) zones have significant negative effects on the proportionate change in median values of owner-occupied housing. While the individual coefficients are not significantly different, if we weight the coefficients by the average share of block groups in each zone (0.30 for AE and 0.05 for AH) there is a significant difference with a p-value for the test of 0.057. Moreover, it is consistent with the higher-risk area having a larger

*Table 12.5   Census-based estimates of Andrew's effects on median housing
            values and rents, 1900–2000*

Independent variable	Home owner values		Rents	
	(1)	(2)*	(3)	(4)**
NOAA / *Miami Herald* damage	−0.014	−0.014	0.071	0.058
	(−0.12)	(−0.12)	(0.58)	(0.49)
Proportion of block group in AE	−0.110	−0.111	−0.085	−0.108
Zone	(−2.61)	(−2.62)	(−2.15)	(−3.21)
Proportion of block group in AH	−0.175	−0.178	0.074	0.020
Zone	(−2.22)	(−2.26)	(0.95)	(0.29)
Proportion of block group in	0.044	0.052	−0.076	−0.006
X500 Zone	(0.64)	(0.75)	(−1.15)	(−0.10)
Constant	0.349	0.342	0.237	0.260
	(14.43)	(13.78)	(10.30)	(13.07)
Other controls for attributes	No	Yes	No	Yes
No. of observations	945	945	977	803
R²	0.012	0.017	0.008	0.041

Notes: The numbers in parentheses are t-ratios for the null hypothesis of no association.
* The controls for the homeowners' equation include the change in the proportion of
owner-occupied homes with five rooms or less, the change in the proportion of one-family
homes, the change in the proportion of owner-occupied mobile homes, the change in the
proportion of homes with two or more bedrooms, and the change in the proportion of
homes with complete kitchens.
** The controls for the rental equation include the change in the proportion of rental
mobile homes and the change in the proportion of rental units with two or more bedrooms.

effect on median property values. In the case of rents, the information only
appears to have an effect in reducing rents for the high-risk areas.

Table 12.6 describes how the effects of damage and areas in risk zones
influence changes in the proportion of the homes in each home value interval and in each rent category. These groupings offer another strategy for
controlling for all the housing attributes that could have been included in
the models in Table 12.5 using medians. The distributional approach may
well offer a better control because of the likely narrower difference in attributes across the housing units in each cell.

These results isolate the increase in the proportions in the low home value
and low rent categories with damage, as we discussed in the case of the
simple models presented earlier, though the significance of these effects is
somewhat lower. The high-risk FEMA zones tend to reduce the number of
homes in the middle value group and increase those in the lowest housing
value group. Interestingly, as our simpler models and analysis of changes

*Table 12.6   Census-based estimates of Andrew's effect on distribution of
housing values and rents*

	NOAA / Miami Herald damage	FEMA Flood Zones		
		Zone AE	Zone AH	Zone X500
Homeowner Values				
Proportion HV<40K	−0.04	0.06	0.04	0.02
	(−0.80)	(3.78)	(1.24)	(0.91)
Proportion 40K < HV < 100K	0.18	−0.01	0.00	−0.08
	(1.89)	(−0.40)	(0.06)	(−1.49)
Proportion 100K<HV<250K	−0.13	−0.06	−0.07	0.07
	(−1.59)	(−2.02)	(−1.32)	(1.52)
Proportion 250K<HV<400K	−0.00	−0.01	0.02	−0.01
	(−0.11)	(−0.60)	(0.99)	(−0.90)
Proportion 400K<HV<500K	0.00	0.01	0.01	−0.00
	(0.03)	(2.52)	(0.72)	(−0.54)
Proportion HV>500K	−0.00	0.00	0.01	0.00
	(−0.23)	(0.82)	(0.59)	(0.05)
Rents				
Proportion Rent<250	−0.05	0.00	0.02	0.05
	(−1.54)	(0.29)	(0.80)	(2.71)
Proportion 250<Rent<500	0.31	−0.08	0.02	0.00
	(4.74)	(−4.05)	(0.42)	(0.11)
Proportion 500<Rent<750	−0.21	0.06	−0.10	−0.04
	(−2.60)	(2.58)	(−2.01)	(−0.94)
Proportion 750<Rent<1000	−0.11	0.00	−0.01	0.02
	(−1.79)	(0.19)	(−0.18)	(0.60)
Proportion Rent>1000	0.06	0.01	0.07	−0.04
	(1.16)	(0.66)	(2.01)	(−1.22)

Note: The numbers in parentheses are t-ratios for the null hypothesis of no association.

in income distributions seem to imply, there is an increase in the proportion of homes in the $400 000 to $500 000 home value group. The change in the composition of the distribution from middle values to the lowest value category seems to dominate the effect at the highest end of the distribution in yielding our results with an overall negative effect for the risk measures with models based on changes in the medians.

To the extent that our hurricane example is representative, the analysis of changes in medians based on spatial areas would best be interpreted as tests of the effect of an amenity (or disamenity) rather than as estimates of the magnitude of its incremental value. As our distributional analysis suggests, there are simply too many changes in the composition of the homes (or rental units) that can be taking place. This is especially true for a large-scale event. With smaller sources of impact the discrepancies may be smaller, but in these cases the ability to detect them reliably may also be diminished.

IMPLICATIONS

This analysis has implications for two areas. The first involves insights into who adjusts to large-scale disasters. Most of differences in adjustment across groups differing in educational and racial background are likely to be due to their economic capacity to undertake changes in their residential locations. The pattern no doubt reflects differences in these groups' available income and wealth. Whites may have access to greater resources to permit their adaptation than Hispanics or black households who do not own their homes.

Several authors' concerns about the confounding effects of household heterogeneity for the composition of communities after exogenous changes in amenities are confirmed with the large-scale damage associated with Andrew. Our analysis of the distributions of income, housing values and rents indicate that the underlying shifts in these distributions can yield ambiguous results for the medians. Nonetheless, the pattern of change in each of these distributions is consistent with the low-income groups being least able to adjust to natural disasters. It seems reasonable to conclude that this lack of responsiveness is due largely to economic capacity and not ethnic influences. Indeed, the hypothesis that the social interaction effect associated with attachment to neighborhoods with 'like' groups was only supported for the highest-income households. All other groups' patterns of adjustment implied movement away from areas with a large fraction of 'their group' in 1990.

Second, our analysis of Census-based estimates of market capitalization of the signals provided by Hurricane Andrew's risk information confirms, in qualitative terms, that Andrew appears to have caused a reconsideration of the risk designation implied by FEMA's flood hazard areas. This finding is also confirmed by a micro-level repeat sales analysis of the hurricane's effect that we report in separate research (see Carbone et al., 2006). A comparison of these two efforts confirms what our analysis of the distribution of homes by home value implies. The composition of impacts of large-scale exogenous events on housing markets is an important part of 'the story'. Analysis of medians or means may offer a credible gauge of relative impact of these exogenous events but it is unlikely to be a reliable basis for estimates of the marginal values or response elasticities.

The contrast between the potential interpretations for the variables we used to control for the damage and risk information associated with the storm also highlights the challenges in using summary statistics to implement a quasi-experimental design. Sorting models imply that heterogeneity in preferences and unobservable features of constraints can have large effects on the adjustments households can make in response to large, exogenous shocks. These differences are likely to show up as changes in the distribution of housing values that may not be easily detected with measures of changes in the central tendency.

Our findings of consistent qualitative results between our analyses of the micro-level repeat sales outcomes and the changes between the 1990 and 2000 Census medians may reflect the long-term nature of the change in perceptions of the hazards of coastal locations. The changes in the distributions of housing values suggest that the highest-income groups appear to be self-protecting and insuring. For the other groups the results suggest that their actions depend on whether they have the economic capacity to adjust their locations.

It is difficult to draw transferable lessons from one analysis of adjustment to a large-scale disaster for other disastrous events, both natural and man-made. It is clear that the economic circumstances of households seems to be the most important factor in understanding responses. Perhaps the primary lesson is that there is a great deal that can be recovered from *ex post* studies of asset prices. A better mapping of the spatial effects of the disasters to residential and other sales prices together with more explicit treatment of neighborhood features offers a research strategy with considerable potential. It will require more detailed and immediate record-keeping after disasters. Our analysis was possible because there was a controversy after Andrew over the performance of Dade County's building inspectors. This public outrage led to a special study of the features of damaged areas and the records that permit our study. More systematic record-keeping creates opportunities to learn how to improve the public role in *ex post* adjustment. This lesson is directly relevant to the challenges facing the Gulf Coast recovery from Hurricane Katrina. Ideally, *ex post* palliative spending will be accompanied by record-keeping that tracks the temporal and spatial dimensions of public support so that it is possible to judge in 2010 what worked and what did not.

NOTES

1. See Cohen and Werker (2005) for an interesting conceptual analysis describing how political and cultural factors can influence the character of efficient public action.

2. Robert Hartwig, Senior Vice-President and Chief Economist of the Insurance Information Institute, used this characterization in describing the impact of hurricanes on economic activity in hurricane-prone counties. He observed that: 'Hurricane Andrew, until September 11, 2001, was the global insurance industry's event of record. For nearly a decade it was the disaster against which all other disasters worldwide were compared . . . Andrew struck Florida in August 1992 with 140 mile-per-hour winds and produced insured losses of $15.5 billion – about $20 billion in current (2001) dollars. . . . Although Andrew has now been eclipsed as the largest insurance event in world history [by September 11] . . . It remains the largest natural disaster on record in terms of insured losses, not only in the United States but world-wide' (Hartwig, 2002, pp. 1–2).
3. Recently, quasi-experiments (such as the one performed here) have raised questions about the viability of earlier hedonic studies on the value of air quality improvements and superfund clean-up (see Chay and Greenstone, 2005; Greenstone and Gallagher, 2005). The current chapter bears on this discussion as well. To the extent that changes in these other environmental amenities result in large distributional changes like those found here, quasi-experiments based on summary data may be biased.
4. This facility was closed after the hurricane due to extensive damage.
5. (Getter, 1992).
6. Statistical models for changes in the proportion of white homeowners do not support a significant positive relationship with damage.
7. When the variable being summarized is a count we apply the fraction of the 2000 block group, that is, from the original 1990 definition. Assuming uniform density of the relevant population in each 2000 block group this process assigns the correct weight to each component. For continuous measures, such as the median income or the median value for homeowners' reports for their home's sale price, the appropriate weight is the fraction of the 1990 block group that is in the 2000 block group. These weights sum to unity when we collapse the 2000 summary statistics to the 1990 map for block groups.
8. In interpreting estimates it might seem implausible to have all negative estimates. However, both the numerator and the denominator in each ratio for each educational category are changing between Census years. Moreover we are not including all educational categories in the decomposition.
9. Breen and Gupta (1997) and Banzhaf and Walsh (2004) find comparable results for Hispanic populations' responses to other sources of environmental risks.
10. $\bar{m}_t$ designates median income in $t = 1990$ and $t + 10 = 2000$; *NOAA Damage* is the average proportion uninhabitable; *AE*, *AH* and *X500* are the proportion of area in each block group in the specific flood zone designation.

REFERENCES

Banzhaf, H. Spencer and Randall P. Walsh (2004), 'Testing for environmental gentrification: migratory responses to changes in environmental quality', Paper presented to the 2004 AERE Workshop, Estes Park, CO, May.

Breen, Vicki and Francis Gupta (1997), 'Coming to the nuisance or going to the barrios? A longitudinal analysis of environmental justice claims', *Ecology Law Quarterly*, **24**, 3–56.

Brock, William and Steven Durlauf (2001), 'Discrete choice with social interactions', *Review of Economic Studies*, **68** (April), 235–60.

Carbone, Jared C., Daniel G. Hallstrom and V. Kerry Smith (2006), 'Can natural experiments measure behavioral reponses to environmental risk?' *Environmental and Resource Economics*, **33** (2), 273–97.

Chay, Kenneth and Michael Greenstone (2005), 'Does air quality matter? Evidence from the housing market', *Journal of Political Economy*, **113** (2), 376–424.

Cohen, Charles and Eric Werker (2005), 'The political economy of "natural" disaster', Working Paper, Department of Economics, Harvard University, November.

Coulson, N. Edward and Eric W. Bond (1990), 'A hedonic approach to residential succession', *Review of Economics and Statistics*, **72** (August), 483–44.

Davis, Donald R. and David E. Weinstein (2002), 'Bones, bombs, and break points: the geography of economic activity', *American Economic Review*, **92** (5), 1269–89.

Getter, L. (1992), 'Inspections: a breakdown in the system', *Miami Herald*, 20 December.

Greenstone, Michael and Justin Gallagher (2005), 'Does hazardous waste matter? Evidence from the housing market and the Superfund Program' NBER Working Paper 11790, November.

Hartwig, R.P. (2002), 'Florida case study: economic impacts of business closures in hurricane prone counties', working paper, Insurance Information Institute, June.

Kahn, Matthew E. (2005), 'The death toll from natural disasters: the role of income, geography and institutions', *Review of Economics and Statistics*, **87** (May), 271–84.

Krugman, Paul (1998), 'Space: the final frontier', *Journal of Economic Perspectives*, **12** (Spring), 161–74.

Miguel, Edward and Gerard Roland (2006), 'The long run impact of bombing Vietnam', NBER Working Paper 11954, January.

NOAA (National Ocean and Atmospheric Administration) (1992), 'Less winds, lots of damage', *Miami Herald*, 20 December.

Tiebout, Charles M. (1956), 'A pure theory of local expenditures', *Journal of Political Economy*.

Wakimoto, R.M. and P.G. Black (1994), 'Damage survey of Hurricane Andrew and its relationship to the eyewall', *Bulletin of the American Meteorological Society*, **75** (February), 189–200.

13. Katrina: a Third World catastrophe?

Edward J. Clay

INTRODUCTION

Natural disasters: a global issue

Hurricane Katrina could prove to be a defining event in the way the global community thinks about natural disasters. The conventional wisdom has been that, with development, disaster risk – the combination of hazard probabilities and vulnerabilities – declines. Richer countries invest in preventive measures, engage in risk transfer and evolve more effective public and private responses to natural hazards. Hurricane Katrina dramatically calls in question such assumptions.[1]

We already know from the assessments that have been done so far with reference to Katrina that there was institutional failure at the federal, state and local government levels.[2] Some of the problems that have been exposed suggest parallels with disasters in poorer, so-called developing countries. First, the professional risk assessments pointing to the serious possibility of a catastrophic event did not result in commensurate disaster prevention measures with regard to either regulation of human habitat and business activity or investment levels of storm and flood protection. Second, the public response appears to have been beset with many of the sorts of problems that are to be found in the manuals on disaster risk reduction (DRR) for developing countries. Third, analyses for developing countries typically focus on the links between vulnerability, poverty and social exclusion or marginality. These parallels raise interesting questions. Are we dealing with problems of a general nature, inherent in the way both public institutions and the private sector manage risk and respond to catastrophic events? Are some developing countries, with much more limited resources and exposed to similar categories of hazard, more effective in at least some aspects of disaster prevention or response? Are higher levels of economic development likely to be associated with structural changes in the economy and society that could increase some of the risks associated with extreme, highly improbable natural events? In a highly speculative way, this chapter will consider and contrast evidence from developing counties exposed to extreme riverine flood and coastal hazards, including Bangladesh, India and the Caribbean islands, with Hurricane Katrina and the Netherlands.

Outline

The chapter has two parts. First there is an overview of the literature on both the short-term economic impacts and long-term economic developmental implications of natural disasters focused largely on developing countries. The overall conclusion is that disaster shocks are likely to have significant but relatively manageable macroeconomic effects on larger economies. However, disasters typically impact most severely on poorer disadvantaged groups and areas, and these should be addressed in public disaster risk reduction measures, anticipated in preparedness and recognized to require comparably disproportionate responses. Second, and speculatively, three issues of DRR identified as significant in the Interagency Performance Task Force (IPET) evaluation (USACE, 2006) – resilience, redundancy in safety and the recognition of changing hazard risks – are considered with some examples from developing countries and the Netherlands. The role of information on disaster risks is found to be critical and problematic.

Definitions

There is no universally accepted set of concepts and definitions in DRR, perhaps because this is inherently a multidisciplinary area and different professional groups have their own overlapping but distinct discourses about both physical and social processes. Broadly those adopted in this chapter are reflected in Figure 13.1. Two specific uses should be noted: exposure is distinguished from vulnerability. The significance of this distinction is apparent in considering public regulation of settlement and economic activity. Land use regulation may be used to avoid exposure by restricting access to and use of areas considered at risk. Building regulation may be used to limit vulnerability to storm or earthquake hazard. The overall macroeconomic impact of a disaster is not simply the aggregation of the direct damage and indirect losses caused by the disruption that are usually reported in post-disaster assessments, for two reasons. First, stock and flow impacts cannot be aggregated in this way as they were, for example, in assessment reports for countries impacted by the 26 December 2004 tsunami (TEC, 2006). Second, the anticipated negative indirect impacts usually fail to consider the adjustment process within the wider economy, including for example positively post-disaster reconstruction.

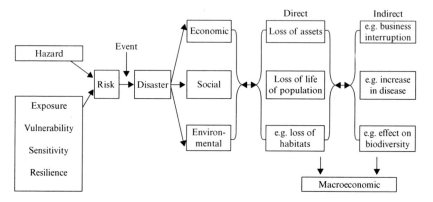

Source: Adapted from Mechler (2005).

Figure 13.1   Natural disaster risk and categories of potential impacts

THE MACROECONOMIC CONSEQUENCES OF NATURAL DISASTERS

Overview

The combined economic damage from Hurricane Katrina, $125 billion, followed quickly by Hurricane Rita, $16 billion, was estimated as over $140 billion, equivalent to more than 1 percent of gross domestic product (GDP) in the world's largest economy. This was clearly a major natural disaster shock from an economic and financial perspective, with potentially significant macroeconomic effects, including fiscal implications. The combined level of insured losses has considerable implications for the global insurance market. What does economic analysis of other large or more extreme disaster shocks – that is, relative to the affected economies – suggest about the likely impacts? The balance of evidence for disaster shocks in developing countries is that the broad macro impact might be a temporary reduction in GDP growth of up to 5 percent. However, this effect is likely to be dampened by the adjustment process within the wider economy, and short-lived. But the localized consequences of Katrina might include lower regional economic growth if there were a reduction in expected returns to investment in New Orleans and the immediate coastal region. The consequences of such a disaster almost invariably are most severe for poorer, socially and economically disadvantaged groups. This section summarizes the empirical evidence and analytical arguments that underlie such broad-brush conclusions.

Disasters and Economic Performance

There is an emerging near consensus, reflected for example in an IMF (2003) review, that natural disasters impact negatively on economic growth, but disproportionately on poorer income groups and spatially on poor communities, and so are a threat to development, including the achievement of international Millennium Development Goals. These conclusions are based largely on an accumulation of empirical evidence from case studies for geographically extensive individual events such as the sub-Saharan Africa droughts of 1991–92 and 1994–95, the 1997–98 '1 in 50 years' El Niño event that had a global footprint, and country reviews for many smaller and micro-economies that have been repeatedly affected by disasters. Mostly, the studies have adopted partial techniques of analysis which are employed an eclectic way.

Macroeconomic aggregates: in trying to establish the impact of disasters for national economies, Benson and Clay (1998), IMF (2003), ECLAC (2003) estimated the effects of a disaster shock, firstly from a comparison of actual year-to-year rates of change, and secondly from a comparison of post-shock short- and medium-term trends (w) in economic aggregates with pre-disaster trends (w_0). This approach provides only an approximate estimate of the likely impact of each disaster shock through time-series and cross-sectional comparison.

More systematic attempts to determine the effects of disaster shocks are hampered by the widely recognized weaknesses of historical data, at least prior to 1990, and quantitative results should be used with caution. For example, Raddatz (2005) finds that only a small part of volatility in macro aggregates of low-income countries is attributable to disasters and humanitarian crises treated as bivariate, dummy variables (23 percent), and these are on average associated with a 2–4 percent reduction in GDP.[3]

After a careful inspection of cases, and allowing for other influences, a number of generalizations have been made from this body of case study evidence, complemented by the results of cross-sectional analyses.

Disasters have a discernible impact on macroeconomic performance. The International Monetary Fund (IMF) (2003) found in a set of case studies that large natural disasters[4] had typically reduced GDP growth by more than 5 percent before recovery returned growth to pre-shock levels.

Exposure emerges as a key explanatory variable of the impact of disaster shocks, associated with the differentiated effects of both climatic and geophysical hazards. Broadly, the greater the exposure, the more likely is a disaster to have negative impacts on short-term economic performance and economic development.

Shocks caused by climatic hazards especially, and those which are more likely to have wider impacts on economic activity such as drought and river basin floods, have a discernible and statistically significant negative impact.

Geophysical hazards such as an earthquake, volcanic eruption or tsunami are more usually associated with localized loss of life, direct damage and disruption, even when on a catastrophic scale. The main recent exceptions have included small (Maldives) and especially micro (Montserrat) island economies.

Seismic hazards and megacities: there is a growing risk of a disaster of internationally catastrophic proportions to many rapidly expanding urban centres including several megacities (Kreimer et al., 2003). Physical geography may have encouraged the location of many such centres in hazardous locations associated with active faults such as oases in arid areas (Iran), and deepwater ports (the Bosphorus and Tokyo–Yokohama).[5]

Small and particularly micro-economies are extremely sensitive to disaster shocks that are more likely to inflict damage and disruption across the economy.[6] Most small island economies are a high-risk category, and the process of climate change and associated rising sea levels is increasing the risk from tropical storms.

Countries exhibit changing patterns of sensitivity over time to broadly similar hazards measured in physical terms (for example extreme rainfall or storm intensity) that are associated with changes in economic structure such as diversification, and also public and private sector actions directed towards DRR.[7]

Disasters are likely to have an asymmetric impact on poverty, more serious for the poor, elderly, disabled, women and households with less flexible income-earning possibilities. Poorer groups are disproportionately living and working in more exposed locations. Poor fishing and handicraft households, dependent on highly specific productive assets, may be more vulnerable than unskilled, migrant labour that can find employment in reconstruction. Behind the movements in broad macroeconomic aggregates, there are likely to be considerable and complex income and asset redistributional effects of disaster shocks. However most economic loss assessments have been poverty-blind, and so much of the evidence is inferred from subsectors affected by the shock. This deficiency is being remedied, with socioeconomic assessment becoming a more normal part of larger inter-agency assessments and efforts made to adopt a standardized approach.[8]

Disaster shocks are likely to put more severe short-term pressures on public finances at provincial (state) and national levels. The empirical evidence, usually derived from the *ex post* inspection of public accounting data, is mixed. The sources of evidence are a limited number of systematic country case studies and public expenditure reviews. The combination of increased expenditure on relief and rehabilitation transfers, combined with downward pressures on revenue, is widely documented. The unplanned real-locations of funding are normal and pervasive and have a potentially depressive impact on public investment in physical and human capital (see the section below). Where there is little or inadequate provision even for rel-atively frequent and predictable forms of disaster such as localized flooding and tropical storms, then such reallocations may become an almost normal element of public finance. These can undermine attempts to move to more coherent sustained strategies for development and poverty reduction.[9] Evaluations of disaster-related aid funding are concluding that reallocation is suboptimal and separate funding is usually appropriate (for example World Bank, 2006).

Fiscal management: the accumulative case study evidence suggests that following a disaster shock the management of the public finances is a key area. In cases where a disaster shock triggered economic destabilization, loss of control over the public finances was a critical factor.[10] Recognition of the potentially disruptive effects of disaster shocks has galvanized inter-national and regional finance institutions to make specific provisions for responding to short-term financing requirements and medium-term recon-struction costs.

Longer-Term Developmental Dimensions of Disaster Risk

State of the debate
The analytical challenge of assessing the longer-term developmental conse-quences of individual disaster shocks or the cumulative effect of a sequence of shocks has not yet been satisfactorily addressed. There is an opportunity to elaborate a fuller theoretical perspective, especially as a special case of the impact of exogenous shocks on longer-term growth. The available evidence includes an accumulation of case study experiences that employ a variety of methodologies and a few attempts at cross-sectional inter-country statisti-cal analysis.[11] There is evidence that DRR measures have had a high rate of return or benefit–cost ratio in highly disaster-prone countries such as Bangladesh and China for which calculations have been attempted. Various official reviews (for example IMF, 2003; DFID, 2006; World Bank, 2004) conclude that this body of evidence is convincing and justifies assigning a higher priority to DRR and the inclusion of DRR within country develop-ment plans such as the poverty reduction strategies being adopted by most low-income countries. However, there are some dissenting, sceptical voices also concluding, on the basis of empirical evidence, that in many cases the socio-economic effects of natural disasters are localized and without long-term macroeconomic significance (for example Albala-Bertrand, 2006).

Economic adjustment and disaster shocks[12]

The literature on natural disasters is unclear on whether the process of economic adjustment amplifies or dampens the direct impact of a disaster shock or series of shocks. Yet this is a critical issue in seeking to understand the longer-run implications of disaster shocks and the relationships between disasters and development. Most of the quantitative evidence on impacts is derived from initial post-event assessments of direct damage to capital and anticipated indirect disruption losses (Figure 13.1). Further retrospective analysis is obliged to take into account purposive responses including official transfers. Nevertheless, there is sufficient evidence to conjecture that the relationships between disaster exposure and vulnerability and development are complex and subject to multiple influences that are also changing. The empirical evidence provides a basis for setting out the issues in a more formal way, suggesting that the outcomes are likely to be context-specific.

First, it will be useful to distinguish:

- disasters that impact relatively more severely and destructively on physical and human capital (geophysical hazards and in some contexts floods and tropical storms); and
- disasters which cause negative productivity shocks, especially on agriculture (not just drought but rainfall and temperature variability and extremes and, in some circumstances, floods and tropical storms).

This distinction also has further potentially important implications because almost all developing countries are experiencing some increased exposure to the first category of shock.[13] Many developed countries may also have growing exposure because of the interactions of spatial development such as coastal residential and tourist development and climate change (Stern Review, 2006). The latter conditions are more likely to be pervasive in low-income countries, especially in sub-Saharan Africa.

Second, depending on exposure to disaster shock, these effects may be considered in terms of a range of economies simplified to:

- a shock that typically occurs at a provincial or state scale, and the adjustment process within the wider national economy, as in the cases of drought or earthquakes in spatially large economies such as Brazil, China, India, the Philippines and the USA;
- shocks to smaller, open economies in which exposure is more economy-wide and so adjustment is internationally based, such as the Caribbean island states.

The ways in which the process of economic adjustment may amplify or dampen the direct impact of a disaster shock (or series of shocks) can be considered first within the alternative growth frameworks that may be employed in modelling development.

Exogenous growth: in a neo-classical world the economic system dampens the effect, as agents adjust their behaviour. It also diffuses the effect, as prices change. For example, the costs of a negative productivity shock to agriculture in one province are diffused within the national economy and at a country level shocks are borne worldwide via price and terms-of-trade changes.[14] However, in the Solow–Swan growth model there is amplification not dampening, as a negative productivity shock causes a less than proportionate fall in steady-state capital and therefore a more than proportionate fall in steady-state output. But if saving and consumption were being inter-temporally optimized then the impact on the present value of utility is dampened. For example, Hallegatte and Horcade (forthcoming) use such a model to explore the dynamic consequences of increasing the levels of severity of disaster shocks (in terms of sensitivity, since exposure cannot be distinguished in a single-sector model). Their results suggest a provocative conclusion, that there is some level of shock beyond which a sequence of disasters results in positive feedback, amplifying the initial reduction in growth.

Endogenous growth implies that investment (in human or physical capital, denoted as H and K below) is likely to be productivity-augmenting (Aghion et al., 1998). So the likely effects of a disaster shock on the rate of return to investment will be influenced by:

- the relative severity of impact on capital and labour;
- the effects on relative prices of inputs into the production of capital;
- the implications of different sectoral capital labour ratios; and
- interactions with market failures.

Such a framework could be relevant to understanding the adjustment process at a state or intra-state level within the US economy to an intense but localized disaster shock such as Hurricane Katrina.

First, is the shock labour- or capital- 'disaugmenting'? To the extent that it destroys buildings and structures, the shock is capital-disaugmenting, reducing the rate of return and the incentive to invest. The differential pattern of investment amongst the smaller Anglophone Caribbean island states is consistent with this effect, suggesting an interaction of size and disaster-type influences on longer-term development.[15]

The loss of human capital (H) is neglected in the disasters literature, but damage to housing, and to infrastructure including schools and health facilities, as well as temporary disruption to business and lay-offs, may encourage a flight of H. The highly educated and skilled are likely to be able to migrate legally and find employment and have the contacts and access to finance to enable them to relocate.[16] If the productivity of stock of H embodies learning by doing, then replacement may be associated with lower productivity[17] and a loss of competitiveness in tradables.

Second, the shock may alter the relative prices of inputs used in the production of physical or human capital. For example the erosion of the tax base may increase the shadow cost of funding teachers and other school-related expenditure.

Third, trade theory suggests deriving changes in factor returns from sectoral impacts. Where the shock impacts primarily on K- and H-intensive sectors reduces the real return to capital and/or skilled labour then adjustment amplifies the effect of the shock (Stolper-Samuelson model type shock).[18]

Where the main impact of the shock is in agriculture that is not K-intensive, then this reduces the returns to land and/or unskilled labour and raises the real return to capital and/or skilled labour. This means that adjustment dampens the effect of the shock; and an example in which the shock is actually good for growth could be constructed.[19]

Fourthly, how does the interaction between shocks and existing market failures change returns to investment? Disaster shocks:

- destroy collateral, thereby worsening credit market failures and reducing investment;
- are a source of volatility and may worsen insurance market failure – examples are to be found from both micro (for example health insurance) and macro (for example overall volatility reduces the attractiveness of alternative locations for investments); and
- increase the scarcity of goods that are subject to 'the tragedy of the commons'.

There is a widespread presumption that development is likely to be associated with a relative reduction in the impact of natural disasters. A more complex multi-sectoral economy has lower levels of exposure, and is assumed to be more resilient. Because of better information, credit and insurance markets that are more complete, there is a higher level of risk-spreading and more optimal investment in risk reduction. The presumption that development will necessarily lead to reduced exposure and vulnerability through regulation of land use and standards in construction needs to be qualified in two ways. There is evidence (for example cautionary examples of the Bay of Naples and the US Gulf Coast) leading us to be concerned that exposure to relatively capital-disaugmenting disasters may be increasing. This may because of private disregard for risks subjectively regarded as very low. Another influence could be moral hazard, for example, because of the provision of federal flood insurance in the USA.

A second long-term issue is the implication of all forms of climate-related hazards increasing in frequency and severity because of climate change (for example Stern Review, 2006). The development implications seem more adverse for two categories of a developing economy: first, low-income countries most vulnerable to agricultural productivity shocks, largely in SSA; and, second, small island states where recent DRR gains are threatened. However the changing risks and vulnerabilities are likely to be so context-specific that a careful review of climate change, the DRR and development strategies is justified at a country level and provincial scale in larger economies.

To summarize, many developing countries are highly disaster-prone in terms of their exposure to extreme physical hazards. The cumulative evidence from recent investigations is that there are also subsets of these countries whose economies are fragile, so that disaster shocks have affected their development trajectories and represent a potential threat to poverty reduction strategies and sustainable development. Furthermore, the high degree of context-specificity, including the dynamics of risk, imply the need for regular reassessment and ensuring that the effects of disasters are better understood at macro and micro scales.

The consequences of a disaster are almost invariably most severe for poorer, socially and economically disadvantaged groups, and this is what happened in New Orleans (Logan, 2008; Pastor et al., 2006). This asymmetry of impacts has profound DRR implications. Also, in preparing against future disasters, extraordinary compensating measures to be focused on those most vulnerable will be required, which implies advanced planning about how this should be done.

ISSUES IN DRR: RESILIENCE, REDUNDANCY AND THE RISK RECOGNITION

Overview

The IPET performance evaluation amongst its many thought-provoking conclusions has highlighted three issues: redundancy with a disaster protection system, lack of resilience within the affected economy and society, and the problem that institutions responsible for providing protection have had difficulty in recognizing and then taking into account changing risks (USACE, 2006). These are issues on which it is interesting to consider possible parallels with developing-country experiences. Are these context-specific local and US issues, or generic problems of DRR? Why is

Dhaka, for example, seemingly more resilient than New Orleans? Why was the Hurricane Katrina death toll much higher than during recent catastrophic floods in Bangladesh, where many more people were severely affected? Why is it difficult to ensure higher safety levels, through redundancy within the flood protection 'system'? Why do institutions responsible for flood protection sometimes fail to recognize an accumulating body of scientific evidence on environmental changes that influence underlying hazard risks?

Resilience: a Candle with Matches

In Bangladesh the extreme riverine floods of 1998 were reported to have affected 15 million, but killed only 140 people. The floods of June 2004 affected 36 million out of 140 million, killed 730 and resulted in damage of $7 billion. Hurricane Katrina killed 1322, affected 500 000 people and caused damage of $125 million.[20]

Some developing countries appear to be resilient in the face of extreme natural hazards. This is partly because these events are expected to occur, and the warnings about actual hazards are getting better. People and governments are inured to minor systems failure and lesser disasters as a feature of life. Travellers to South Asia will be familiar with there being a candle with matches in almost every hotel room. Some, including the author, regularly avoid the use of lifts in hotels and public buildings in these countries because of the risk of a power cut. In the bustees, or shanty towns, of Bangladesh people have learnt to cope with recurrent flooding. The techniques of storage within bamboo-framed houses allow possessions to be kept well above floor level, so that households can cope, or better survive, with 1–2 ft of flood water in their homes, workshops and small retail outlets. If the flooding becomes more extreme, then they may be forced out with their meagre possessions and flee, seeking higher ground or refuges on foot or by boat, bullock cart or rickshaw. The less complex technologies which involve some degree of adaptation to flood, and the very multiplicity of forms of transport and lack of near total dependence on electrical power, are an advantage.[21] Within three months it was scarcely possible to discern in Dhaka any remaining visible impacts of extreme floods.

A possible answer to the 'candle with matches' conundrum is that it implies an acceptance of residual uncertainties which are not completely removed by current flood protection measures. For a business such as a hotel or more seriously a public institution in a high-income, industrialized economy to acknowledge such risks is itself a potential threat to consumer confidence and to potential investor expectation of rates of return. Sometimes it is convenient to act as if there were no problem.

Redundancy within Flood Protection Systems

The deliberate provision of redundancy within a flood protection system or other disaster protection system is uncommon in developing countries or indeed in most high-income developed countries. A striking exception is the Netherlands, and a few context-specific points help to provide a contrasting perspective to developing-country experiences. The evolution of flood protection within the Netherlands makes it possible to maintain already existing structures as secondary lines of defense. But modern planning, notably the Delta Project, explicitly incorporates redundancy into flood and coastal protection to provide secondary lines of defense to minimize the potential impact of catastrophic events (Van de Ven, 1993).

The process of reclamation from the late medieval inundations was inherently piecemeal and compartmentalized. Collective experience is that dikes do burst and are overtopped, so that one generation of dikes is maintained after a new defense line and related drainage are established, as for example in the northern Netherlands. The centrality of flood protection to the Dutch nation and its economy was dramatically demonstrated by the World War II breaches and then the catastrophic 1953 floods, so that high safety levels are understood to necessitate fail-safe backup systems that may not be called into use (Box 13.1). There are now water councils in the Netherlands, separate, democratic, regional-level institutions for water management and protection with tax-raising powers. Perhaps these arrangements encourage a cohesive approach and informed attitudes to flood protection: the Dutch people expect and support the public funding of costly protection assuring one in 1000-year risk levels. Unfortunately few developing countries can afford or even yet recognize the need for such levels of safety, especially in publicly, often aid-funded and centrally organized protection systems.

In South Asia the colonial state displaced the feudal rulers as the providers of public safety, perhaps thereby contributing to a massive problem of moral hazard. The post-independence government has taken on that role and is expected to construct and maintain the embankments and provide relief if there is a disaster. There is no established tradition of local participation in determining needs and solutions, and instead centralized bureaucracies at the national level in Bangladesh and at the state level in India have determined priorities and provide protection, whilst accommodating the demands of political patronage.

Floods are likely if protective structures are overtopped or breached and more frequently because of more localized drainage problems behind the major river embankments and inside protected polders (see Box 13.2). There is recognition of the need for secondary safety, as in rural Bangladesh. This is reflected in kinship groups who live on the flood plains behind the

---

BOX 13.1 SECONDARY SAFETY IN THE NETHERLANDS: THE STORM SURGE BARRIER ON THE HOLLANDSE IJSSEL

The storm surge barriers at Maelant at the mouth of the River Rhine is the final step of the Delta Project launched in response to the 1953 floods. Behind this primary defense line at the entrance to Rotterdam Port is the storm surge barrier at Krimpen near the mouth of the Hollandse Ijssel (a distributary of the River Rhine), the first Delta Project structure completed in 1958. This barrier protects the densely populated central part of the Netherlands known as the Randstad from a potential storm surge up to 1.5 m above that in 1953, assumed as a one in 1000-year event. The barrier was originally constructed as a more cost-effective protective solution than further dike elevation along the Hollandse Ijssel. It is now maintained as the secondary line of defense. There is a risk that the complex Maelant barriers could fail to close in an emergency (a 4% probability). Initially in 1958 there was a single barrier at Krimpen, which was then supplemented by a second barrier. The opening and closing motors are normally powered from the electricity mains grid. There are diesel-powered generators in the event of grid failure. If there were a motor failure, then six men could lower the barriers within 20 minutes and local fire service personnel are always on standby. The storm surge warning from the coast at the Hook of Holland is about 1 hour 25 minutes. The dikes behind the barrier now serve as a tertiary line of defence.

Source: Van de Ven (1993, p. 265)

---

embankments continuing to build their homesteads on artificially elevated mounds. There are also incoherent, 'beggar my neighbour' local flood protection measures. The controversy surrounding the post 1988 disaster Bangladesh Flood Action Plan perhaps marked the beginnings of a genuinely participatory process for flood protection in which civil society became engaged and the debates were increasingly open within a reinvigorated democracy.

Disaster Risk Recognition and DRR Planning

The hurricane protection system for New Orleans was based on design criteria developed in 1965–66. Despite updates in information on hazard risk

the protection system was designed and continued to be developed according to these agreed definitions (ILIT, 2006; USACE, 2006). Following catastrophic failure in 2005 a different, system-wide, risk-based approach is
being actively considered. Some of the contributors to this symposium are
part of that process of reassessing the flood protection needs for New
Orleans and are seeking to draw lessons from DRR more generally. After
examining a number of country cases, and acknowledging the specifics of
every situation, there appear to be at least three closely related and recurrent or interactive features in DRR policy practice that can increase the
likelihood of extreme events having a catastrophic impact. Then these
extreme events become a catalyst for redirections in disaster planning.
These interacting practices are:

- tunnel vision or professional compartmentalization;
- policy-level reluctance to recognize changing levels of risk associated
  with potentially extreme but improbable events; and
- policy-level reluctance to prioritize investments in minimizing
  the impact of potentially extreme but seemingly low-probability
  events.

These practices are illustrated with case studies of flood control in
Bangladesh and volcanic hazards in the Eastern Caribbean.

Tunnel vision
The evolution of riverine flood protection in Bangladesh is summarized in
Box 13.2, suggesting a thought-provoking parallel with the history of protection in New Orleans and south-east Louisiana. The flood protection,
drainage and irrigation plan established in 1964 was remorselessly implemented by what became the Bangladesh Water Development Board
(BWDB), with little regard to mounting evidence of implementation problems or changes in underlying conditions that called into question parts of
the plan. The urgent need for enhanced protection for the emerging megac-
ity of Dhaka was only recognized following the disastrous floods of 1987
and 1988. But the BWDB has continued to envisage a system of river
defenses combined with drainage and irrigation that is presently institu-
tionally if not technically infeasible, and so largely unfunded. Large-scale
internationally funded food-for-work programs had provided the resources
for decentralized, incoherent and widely inconsistent local-level flood protection and drainage, including thousands of miles of embankments,
drainage canals'and unblocked or re-excavated watercourses. The floods of
1998 and 2004 underscore the continuing failures. Meanwhile this rural
economy has adapted to flood risk through progressively substituting

higher-productivity dry season lift-irrigation agriculture for flood-prone monsoon season rice and jute. Bangladesh is developing despite recurrent disasters, if still too slowly to make massive inroads into widespread poverty.

BOX 13.2   FLOOD CONTROL IN BANGLADESH

Following severe floods in 1954 and 1955, the government of Pakistan set up a Flood Commission in 1955, which was followed by a UN technical assistance team (the Krug Mission) in 1956, to examine possible flood control measures. That led to the establishment of the East Pakistan Water and Power Development Authority (EPWAPDA) in 1958, the water development functions of which were allocated to the present Bangladesh Water Development Board (BWDB) after Bangladesh became independent in 1971. With international assistance, EPWAPDA formulated a master plan which comprised 58 projects intended to provide flood protection to 5.8 million hectares. This formed the basis for the major flood control, drainage and irrigation projects undertaken by EPWAPDA/BWDB between the mid-1960s and the 1980s.

In reviewing the EPWAPDA Master Plan in 1966, the World Bank advised against its adoption on the grounds of inadequate availability of data and resources. The World Bank went on to undertake a Land and Water Sector Study which emphasized small, quick-return flood control projects and the expansion of irrigation by means of low-lift pumps and tube-wells. Although this strategy was opposed by the BWDB and was never formally endorsed by the government of Bangladesh, it became the strategy actually implemented through the 1970s and 1980s because of its influence on donor funding.

By 1988, successive governments, supported by donor funding (including food-for-work programs) had built about 7500 km of embankments along both sides of the Teesta and Ganges rivers, the right bank of the Brahmaputra-Jamuna River, about half the left bank of the Jamuna river, long stretches of the Padma and lower Meghna rivers, several distributaries of the main rivers and most eastern rivers, together with embankments enclosing extensive polder project areas on tidal and river floodplains, submersible embankments around some deep basin sites in the north-east, and embankments along long sections of the coast. Those works had absorbed about 10 percent of the country's annual development plan budgets. Despite expenditure, several severe floods occurred in the 1970s and 1980s, culminating in the catastrophic floods in 1987 and 1988 which led to a further surge in government and donor commitment to flood protection made manifest in the Bangladesh Flood Action Plan.

Embankments along the major rivers were built to withstand overtopping in 1 in 50 or 1 in 100-year floods. These standards proved adequate in the 1988 and 1998 floods. Where embankments were damaged, this was caused by riverbank erosion, piping (passage of water through or under embankments) or purposeful cuts in the embankments. Conflicts of interest have arisen between people living inside and outside major river and polder embankments, leading to cuts being made in embankments in attempts – usually futile – to relieve high flood levels outside the embankments or internal drainage. However, in the case of the now largely urban Dhaka Narayanganj Demra Polder the embankments were saved from overtopping in the 1988 and 1998 floods by locally organized emergency measures to raise the height of threatened sections by sandbagging.

Source: Brammer (2004).

Volcanic hazard risks in the Eastern Caribbean
In a globalized economy a very high proportion of private investment is footloose. The studies by the author and colleagues into the economic consequences of disaster risk in the Eastern Caribbean suggest that concerns about how hazard risks might negatively affect investors, especially inward investor sentiments, can inhibit recognition of disaster risks and DRR strategies. The near reckless disregard of volcanic hazard risk in Montserrat prior to the eruption in 1995 (Box 13.3), lack of transparency in Dominica in 1998–99 and, more generally, the underfunding of regional seismic hazard research prior to these events are consistent with this conjecture.

Scientific hazard monitoring and information dissemination have been organized in the Caribbean area at a regional level in ways that reflect colonial history. For seismic-volcanic monitoring, Dominica and Montserrat contributed to and relied on the Seismic Research Unit (SRU), based in Trinidad. There is a convention of extreme caution in making potentially sensitive information available, at least in those former European colonies that have not had a tradition of open government. The centralized authority in colonial times had a general responsibility – a contingent liability, in insurance terms – in the event of a disaster, and non-exclusion of potential beneficiaries is still seriously qualified by administrative reluctance to make information available.

Montserrat   There was widespread understanding within the scientific community by the late 1980s that volcanic hazard posed a serious threat to the people and economy. However, this knowledge made no impact on policy when there were opportunities to reduce exposure and prepare against mounting risks of a major eruption. The SRU also accepted a role of providing information in confidence to the locally elected head of the island council with disastrous and economically wasteful consequences (Box 13.3).

BOX 13.3   MONTSERRAT: IGNORING VOLCANIC HAZARD RISKS, 1989–95

The UK Overseas Territory of Montserrat (pre-eruption population 12 000) is the extreme case of the failure to recognize publicly and to address scientifically well understood and increasing risks of a volcanic eruption. After the devastation of Hurricane Hugo in 1989, the need for relatively comprehensive reconstruction offered an opportunity to relocate key infrastructure and social facilities, including government headquarters, the electricity plant and hospital, to areas of lower volcanic hazard. The risks from an eruption similar to that which had destroyed St Pierre on Martinique in 1903 were widely recognized in the scientific community and a risk hazard diagram based on recent volcanological research had even found its way into a US-published school textbook to show why human activity might have to be relocated in the face of volcanic hazard.

Instead all facilities were rebuilt in high-risk areas in and around the main town and port of Plymouth. Subsequently, initial evidence of increasing seismic activity during 1992–94 did not lead to enhanced monitoring and was not made public or even drawn to the attention of the colonial governor responsible for emergency response. A 1994 disaster emergency response plan acknowledged but did not consider volcanic hazard. The hospital rebuilding was completed in 1994. After the unanticipated eruption began in 1995 all reconstructed facilities had to be abandoned, temporary facilities established and rebuilt again from 1998 onwards in a zone considered safe from the continuing eruption and related pyroclastic flows.

Source: Clay (1999).

Dominica The island is seismically extremely active. The SRU had successfully monitored volcanic alerts in the 1970s and 1980s and initiated risk assessment and risk mapping. However, the handling of a new volcanic alert in 1998–99 raises the difficult but important issue of how scientific information should be disseminated to the wider public to ensure that both public and private sector institutions make rational decisions on natural hazard risk. Despite previous alerts, a volcanic emergency plan had to be specially prepared and emergency exercises were carried out. But little precise information was made available publicly on the nature and extent of the risks posed. There was considerable uncertainty in the private sector and amongst civil society organizations about the precise nature and level of risk posed, how the crisis might evolve, and appropriate responses. This resulted in a confused range of reactions. For example, some insurance companies temporarily stopped taking on new business in the southern part of the island, whilst a few did not renew existing (annual) policies.

These experiences have had a constructive outcome. The funding of seismic hazard monitoring and researching in the region has been enhanced. There is also an official acceptance at the regional and country level of the need to provide information on the evolving status of seismic hazard openly and freely as a public good, for example, on the Seismic Research Unit website and in scientific papers. It is too soon to do more than speculate on the consequences of this new openness as a potential influence on both public and private investment within and amongst the island economies.

There are uncomfortable parallels between the Eastern Caribbean and Hurricane Katrina. In professional circles and the media, New Orleans was also widely discussed and cited as a disaster hotspot – at significant risk to a catastrophic storm (for example Dilley et al., 2005; Van Heerden and Bryan, 2006). However there was a lack of connectivity between professional understanding of the increasing hazard risk and what was actually done at a policy level to address these problems.

CONCLUSIONS

An overview of the literature on both the short-term economic impacts and the long-term economic developmental implications of natural disasters focused largely on developing countries leads to two conclusions directly relevant to this volume on Hurricane Katrina. Disaster shocks are likely to have significant but relatively easily manageable macroeconomic effects on larger economies. However disasters typically impact most severely on poorer disadvantaged groups and areas, and these should be addressed in public disaster risk-reduction measures, anticipated in preparedness and recognized to require comparably disproportionate responses. Endogenous growth theory also provides a useful but yet to be exploited framework for exploring the process of adjustment to a disaster shock at both state or provincial and economy-wide levels.

Economic theory suggests that consumption and investment and other resource allocation decisions are more likely to be optimal when continuously based on fuller information about hazard risks than when abruptly modified by sudden discrete changes in perceived risks after a catastrophe. The implication is that the international community and governments should both support the generation of knowledge on disaster risks and ensure the wide dissemination of information about these risks in an understandable form to stakeholder groups as global and national public goods. On the evidence of recent major disasters including Hurricane Katrina, US federal and state bodies as well as similar institutions in most other countries should go further in accepting and fulfilling this responsibility.

There are several recurring issues concerning effective DRR which concern both the generation and use of information on natural hazards as a public good (Benson and Clay, 2004). The ILIT and IPET evaluations acknowledge the need for supporting research into the dynamics of hazard risk and its disaster potential (ILIT, 2006; USACE, 2006). But in seeking to understand the implications of Hurricane Katrina, if only the professional community of scientists and disaster specialists realized that this was a catastrophe waiting to happen, then the more immediate issue concerns information use.

The use of hazard risk information is unsatisfactory in many cases, especially if participatory processes of DRR are envisaged. This is partly to do with the probabilistic nature of the information generated by scientific research and monitoring. This raises a series of questions:

- What forms of information is it appropriate to make available to various stakeholder groups?
- How can scientific information be disseminated in an easily understandable form?
- How should scientific information be used and with what implications, bearing in mind that it will be probabilistic and so difficult to take into account?
- What role should scientists play in informing the general public and other stakeholders directly about natural hazard risk and uncertainty?
- What are the institutional implications?

There are encouraging examples of improving practice following on from the sequence of recent high-profile catastrophic events, perhaps including Hurricane Katrina. The problem of disaster shocks is being mainstreamed by both international and regional financial institutions and governments (ODI, 2005). The concerns about global climatic change have contributed to ensuring interest and funding on climatic processes. Progress in using information to reduce the impacts of tropical storms is impressive. There is evidence too of institutional learning about the need to ensure wider dissemination of information, such as seismic risk assessments in the Caribbean and tsunami risk in the Indian Ocean region. But the inherent problematic remains of public policy and private expectations being too heavily influenced by immediate events. As the authors of the most comprehensive histories of water management in the Netherlands conclude:

> If we have experienced rare events, these cannot be passed on. After one or two generations people tend to forget a disaster or to ignore its reality and impact when more urgent problems present themselves. (Van de Ven, 1993, p. 290)

The generality of some of the research findings and the range of country examples presented in this chapter also suggests a wider conclusion – the discourse that separates the world simplistically into developing and developed countries or First and Third Worlds is becoming increasingly inappropriate in a globalized economy. This is especially so as the potential effects of climate change are introduced into models and analyses to explore issues of natural disaster risk.

NOTES

1. For example the author of this chapter was not alone in both complacently assuming and stating (on BBC TV), even when the levees protecting New Orleans had broken, that the US with all its human and physical resources would quickly and effectively cope with the immediate human consequences of this catastrophic event. Then it seemed reasonable to assume that public institutions and private sector would begin to ensure a rapid return to 'normality'. Beyond that, the failure of flood protection structures would raise choices about the longer-term strategy for disaster risk management that might not be so readily resolved.
2. ILIT (2006), NRC (2006a; 2006b) and USACE (2006) provide complementary accounts of failures in flood protection, whilst for example Brinkley (2006) and Van Heerden and Bryan (2006) document institutional failures in responding to the threat posed by Hurricane Katrina and its impacts.
3. The categories in Raddatz's (2005) analysis do not discriminate between natural disasters and humanitarian crises in a satisfactory way. Famines, which are frequently triggered by a natural disaster, are treated as humanitarian crises.
4. Events that satisfy one of three criteria: damage equivalent to 0.5 percent GDP or seriously affecting 0.5 percent of the population or deaths equivalent to 1 in 10000.

5. The extreme earthquakes in Japan and Turkey have been absolutely the most costly (damaging) disasters along with hurricanes in the USA.
6. The 26 December 2004 tsunami directly affected 30 percent of households in the Maldives and disrupted, at least temporarily, the whole tourist sector. In contrast only 17 percent of the Thai tourism sector was located in the affected provinces.
7. The greater resilience of most eastern Caribbean states to disasters since 1990 is associated with diversification out of agriculture, as well as measures to protect infrastructure. The decline in severity of the macroeconomic impacts of drought in India since the1960s and of flooding in Bangladesh since the late 1980s are also partly the result of diversification, in effect of economic development.
8. For example, the United Nations (UN) organized a livelihoods assessment based on the ESCAP (2003) methodology following the May 2006 Jogjakarta earthquake to provide the basis for a livelihoods recovery strategy.
9. Benson and Clay (2004) provide case examples such as Dominica in the 1980s and 1990s and Malawi from 1992 to 2001.
10. Examples of destabilization include Bangladesh in 1974, Dominica in 1979, and Malawi in 1992 and 1994. Most recently the Maldives in 2005 came close to destabilization. In contrast, Bangladesh since 1988, and Botswana, Namibia and Zimbabwe in 1992, show the importance of fiscal stabilization, even if there are initially severe costs in terms of the cutbacks in planned programs, including social expenditure and capital investment.
11. Benson (2003) points to an element of circularity in the analysis – in the categorization of countries as higher risk where there have been demonstrable severe impacts.
12. This section draws on a personal communication from Tony Venables about economic adjustment and the impact of climate shocks in the context of the Stern Review (2006).
13. Raddatz (2005) finds a relatively higher incidence of geophysical disasters in middle-income countries. This phenomenon may be a geographical coincidence or reflect the enhanced levels of exposure that are initially associated with urbanization and industrialization and low levels of regulation of land use and buildings.
14. These dampening effects are exemplified by the impacts of tropical storms that occur almost annually in the Philippines archipelago with scarcely discernible impacts in terms of volatility of macroeconomic aggregates. Occasionally more widespread devastation has impacted beyond the Philippines as the largest supplier on international copra market prices, but the effects are dampened by substitution possibilities with palm oil. Albala-Bertrand (2006) in also looking at localized disasters concludes that the initial negative impacts are strongly dampened by the response from within the wider economy.
15. The differential rates of growth and its sectoral composition in Barbados compared with the ECCB islands are consistent with this conjecture. There has been a disproportionately high level of post-independence investment in transport, tourism and services in Barbados. Barbados is also least at risk to hurricanes or indeed seismic hazards amongst these islands (Crowards, 2000).
16. Unfortunately systematic studies of the impacts of disasters on human capital are lacking. Studies of households forced to relocate by US hurricanes suggest that middle-income groups are more likely to relocate voluntarily and less likely to return to higher-risk zones.
17. For example, the Bangladesh railway system was predominantly staffed until independence in 1971 by Urdu-speaking Biharis. They were then excluded from employment and most subsequently migrated to Pakistan. Productivity levels never recovered to pre-1971 for decades, despite investments in re-equipping the railways and training of a replacement workforce.
18. The hi-tech medical research sector in New Orleans disrupted by Hurricane Katrina laid off staff who have migrated to other competitive centres. After a year and slow rates of recovery in the medical sector, there were concerns that New Orleans had lost its competitive edge in this and other skills-intensive hi-tech sectors dependent on institutional clusters of excellence and that the growth trajectory is shifting towards a lower-skill more labour-intensive tourism sector.

19. The diversification out of export crops in the Caribbean into tourism and financial services and outmigration of unskilled labour is consistent with such an effect (Benson and Clay, 2004).
20. CRED/OFDA disasters database (www.cred.be).
21. For example, even gas pumps are electrically powered in most developed countries.

REFERENCES

Aghion, Philip and Peter Howitt (1998), *Endogenous Growth Theory*, Cambridge, MA, MIT Press.

Albala-Bertrand, J.M. (2006), 'Globalization and localization: an economic approach', in E.L. Rodriguez, E. Quarantelli and R. Dynes (eds), *2006 Handbook of Disaster Research*, New York: Springer.

Benson, Charlotte (2003), 'The economy-wide impact of natural disasters in developing countries', London School of Economics, Unpublished doctoral thesis.

Benson, Charlotte and Edward J. Clay (1998), 'The impact of drought on sub-Saharan African economies: a preliminary examination', World Bank Technical Paper 401, Washington, DC: World Bank.

Benson, Charlotte and Edward J. Clay (2004), *Understanding the Economic and Financial Impacts of Natural Disasters*, Disaster Risk Management Series No. 4, Washington, DC: World Bank.

Brammer, Hugh (2004), *Can Bangladesh be protected from floods?* Dhaka: The University Press Ltd.

Brinkley, Douglas (2006), *The Great Deluge*, New York: William Morrow.

Clay, Edward J. (1999), 'An evaluation of HMG's response to the Montserrat volcanic emergency', 2 vols, *Evaluation Report* EV635, London: DFID.

Crowards, T. (2000), 'Comparative vulnerability to natural disasters in the Caribbean', Staff Working Paper No. 1/00, Caribbean Development Bank, St Michael, Barbados.

Dilley, Maxx et al. (2005), *Natural Disaster Hotspots: A Global Risk Analysis*, Washington, DC: World Bank.

Department for International Development (DFID) (2006), 'Reducing the risk of disaster risk: helping to achieve sustainable poverty reduction in a vulnerable world', A DFID Policy Paper, London, March.

ECLAC (2003), *Manual for Estimating the Socio-Economic and Environmental Effects of Natural Disasters*, Quito: United Nations Economic Commission for Latin America and the Caribbean.

Hallegatte, Stephane and Jean Charles Horcade (forthcoming), 'Why economic dynamics matters in assessing climate change damages: an illustration on extreme events', *Environmental Economics*.

IMF (2003), 'Fund assistance for countries facing exogenous shocks', report by the Policy Development and Review Department, Washington, DC.

Independent Levee Investigation Team (ILIT) (2006), 'Investigation of the Performance of the New Orleans Flood Protection Systems in Hurricane Katrina on August 29, 2005', Final Report, Berkeley, CA: University of California.

Kreimer, A., M. Arnold and A. Carlin (eds) (2003), *Building Safer Cities*, Disaster Risk Management Series No. 3, Washington, DC: World Bank.

Logan, John R. (2008), 'Unnatural disaster: social impacts and policy choices after Katrina', this volume, pp. 279–97.

Mechler, Reinhard (2005), 'Cost–benefit analysis of natural disaster risk management in developing countries', Working paper for GTZ.

National Research Council (NRC), Committee on New Orleans Regional Hurricane Protection Projects (2006a), 'Structural performance of the New Orleans hurricane protection system during Hurricane Katrina: letter report', www.nap.edu/catalog/11591.html.

National Research Council (NRC), Committee on New Orleans Regional Hurricane Protection Projects (2006b), 'Second report of the National Academy of Engineering/National Research Council Committee on New Orleans regional hurricane protection', www.nap.edu/catalog/11668.html.

Pastor, Manuel, Robert D. Bullard, James K. Boyce, Alice Fothergill, Rachel Morello-Frosch and Beverly Wright (2006), 'In the wake of the storm: environment, disaster, and race after Katrina', New York: Russell Sage Foundation, http://www.russellsage.org/news/katrinabulletin2.

Raddatz, Claudio (2005), 'Are external shocks responsible for the instability of output in low-income countries?' World Bank Policy Research Working Paper 3680, Washington, DC: World Bank, August.

ODI (2005), 'Aftershocks: natural disaster risk and economic development policy', ODI Briefing Paper, November.

Stern Review (2006), *The Economics of Climate Change*, Final Report, London: HM Treasury, http://www.hm_treasury.gov.uk/independent_reviews/stern_review_economics_climate_change/sternreview_index.cfm.

Tsunami Evaluation Coalition (TEC) (2006), 'Synthesis report', Final draft, London: ALNAP.

US Army Corps of Engineers (USACE), (2006), 'Performance evaluation of the New Orleans and Southeast Louisiana hurricane protection system', draft final report of the Interagency Performance Evaluation Task Force, Volume 1 – 'Executive summary and overview'.

Van de Ven, G.P. (ed.) (1993), *Man-Made Lowlands: History of Water Management and Land Reclamation in the Netherlands*, 3rd edn, Utrecht: Stichting Matrijs.

Van Heerden, Ivor and M. Bryan (2006), *The Storm*, New York: Viking.

World Bank (2004), 'Natural disasters: counting the cost', Washington, DC, http://web.worldbank.org/WBSITE/EXTERNAL/NEWS/0,,contentMDK:20169861~menuPK:34458~pagePK:64003015~piPK:64003012~theSitePK:4607,00.html

World Bank (2006), *Hazards of Nature, Risks to Development: An IEG Evaluation of World Bank Assistance for Natural Disasters*, Washington, DC: Independent Evaluation Group.

14. Hurricane Katrina and housing: devastation, possibilities and prospects

Raphael W. Bostic and Danielle Molaison

INTRODUCTION

On the morning of 29 August 2005, Hurricane Katrina, which in five days had grown from a tropical storm to a major category 5 hurricane with winds exceeding 172 miles per hour, made landfall near Buras, Louisiana and over the next two days roared through southern and central Mississippi until dissipating in the Midwestern United States.[1] When the storm's eye made its closest approach to downtown New Orleans (23 miles to the east), the sustained winds over the metropolitan area likely remained weaker than category 3 strength. Despite this weakening at landfall, due to its huge size and the category 5-strength waves generated the day before landfall, the hurricane produced a massive storm surge along the northern Gulf Coast.

Hurricane Katrina ravaged nearly 93 000 square miles in Louisiana, Mississippi and Alabama.[2] Over 700 000 residents of the Gulf Coast were acutely impacted by the significant structural damage and flooding wrought by the storm. Of these, over 90 percent were Louisianans.[3] Actual storm damage varied with location relative to the storm's eye. Those located east of the eye (commonly referred to as the 'dirty side' of a storm) suffered extreme structural wind damage in addition to the storm surge. Other areas were subjected mostly to extended periods of heavy rains and the storm surge, both of which resulted in flooding.

Consequently, most of the damage in Louisiana was related to flooding while Mississippi suffered mostly wind damage. At least 633 950 residents in Southeast Louisiana were displaced as a result of flooding only.[4] When compared to Mississippi, Louisiana received relatively little structural wind damage from the hurricane. The Congressional Research Service estimated that 2384 residents of Louisiana suffered catastrophic structural wind damage, and only 161 lived in New Orleans. Conversely, over half of the

65 996 residents affected in Mississippi suffered complete obliteration of entire neighborhoods due to wind damage.[5]

In New Orleans, the storm surge overtopped large section of levees east of New Orleans and breached the levee system at several points within the city. Overall, about 80 percent of the city was flooded to varying depths up to about 20 feet, within a day or so of the eye's landfall.[6] Importantly, if not for levee failures, the city would have largely been spared from the catastrophic flooding that ensued.

Over 200 000 structures within the city limits were damaged as a result of Katrina.[7] While a small number of taller structures on the high ground suffered wind damage only, most of the damage was associated with extreme flooding due to the failure of the levee system. Many buildings were mired in several feet of water until 11 October 2005 – 43 days after Katrina – at which time the Army Corps of Engineers reported that all floodwaters had been removed from the city.[8] Even now, many neighborhoods remain in a state of disrepair.

With 66 609 damaged owner-occupied units and 67 735 damaged rental units, the housing sector accounts for more than half of these structures.[9] Thus, much of the future of the city depends on decisions made regarding the disposition of this housing stock. This chapter considers the issue of New Orleans' housing stock by detailing Hurricane Katrina's effect on the stock and assessing alternatives for reconstruction. It first provides some background on key aspects of New Orleans and how it developed. These underlie much of what transpired and are important for considering how to move forward. Next, the damage to the housing stock is described. After documenting recovery efforts to date, the chapter then turns to the issue of rebuilding, and identifies several straightforward alternative approaches. These vary in terms of the scope of redevelopment and the extent of public assistance to be offered. We conclude with some final thoughts regarding this urban catastrophe.

BACKGROUND

The Geography of New Orleans

The Greater New Orleans area is made up of ten parishes as shown in Figure 14.1. The population for the area was nearly 1.5 million in July 2005, pre-Katrina.[10] Orleans Parish marks the official boundary of the city of New Orleans, which had a population of 458 393 before Katrina hit.

Elevation has long been an issue for those settling in New Orleans. The oldest areas of the city, which contain the greatest concentration of historic

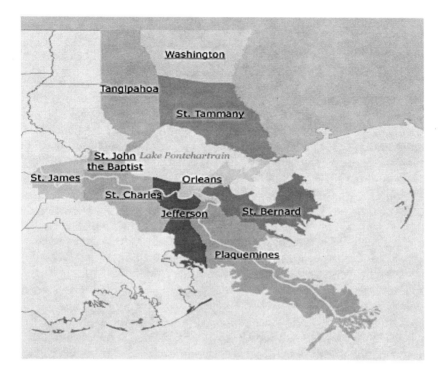

Source: Greater New Orleans Community Data Center, http://www.gnocdc.org.

Figure 14.1 The parishes of Greater New Orleans

structures, are located on the highest ground along the Mississippi River. These neighborhoods are located on a natural levee formed over hundreds of years as the river flooded each spring and deposited sediment along its banks. Structures located in these areas, which include the French Quarter, the Garden District and the Central Business District, were least susceptible to flooding and water damage.

The lower-lying areas in New Orleans were not substantially built out until the 1920s, when development began to extend north toward Lake Pontchartrain. This 'backswamp' area, which includes the Mid-City, Gentilly and Lakeview districts, was developed after the technology to drain this land was discovered. The next expansion of the developed city, New Orleans East, took place in the 1960s and 1970s, also on drained and filled marshland.

Elevation shaped the decisions that builders made in early years. The oldest homes in New Orleans, built from the late eighteenth century until the mid-nineteenth century, are found in the neighborhoods first settled during the colonial period. The Creole Cottage and Creole Townhouse are found in the French Quarter, Marigny and Bywater neighborhoods along the Mississippi River.[11] These housing types were typically set at or near ground level, but they are at some of the highest elevations in the city.

As settlement in New Orleans expanded upriver and on lower-lying land, the Raised Centerhall Cottage and Double Gallery House also became common housing types until the mid-to-late nineteenth century. The predominant New Orleans housing type, the Shotgun House, was built from 1850 until 1910. It is found in all historic neighborhoods of the city. Housing in the backswamp fill areas was often California-style Bungalow, built from the 1920s until the mid-twentieth century. These types were all typically raised above the ground on brick piers.[12]

By contrast, most of the housing in the New Orleans East section of the city is slab-on-grade construction, built without regard to its low ground elevation. Designed as a master-planned community, the original developers abandoned the project after going bankrupt with the 1980s oil crash.[13] This neighborhood, totaling 25 percent of the land area in Orleans Parish, thus became an ad hoc assembly of structures indistinguishable from other American suburbs built during the same time period.

Figure 14.2 depicts the concentration of housing units by age within the city. Neighborhoods with 81 to 100 percent of the homes built before 1950, such as those in the French Quarter, are shaded black. The shade of grey becomes lighter as the percentage of older homes for the neighborhood declines. Areas such as New Orleans East, for example, contain few, if any, housing units built before 1950. Overall, 45 percent of the housing in New Orleans, totaling 129 376 units, was built before 1950.[14]

Damage

The damage to New Orleans originated from two sources. The massive storm surge overwhelmed the levees of the Mississippi River Gulf Outlet Channel and flooded New Orleans East and St Bernard Parish. Moreover, on the day of the storm, Lake Pontchartrain flooded 80 percent of the city when levees failed at multiple points along the city's canal system. As noted earlier, over 200 000 structures within the city limits were damaged.

Hurricane Katrina imposed both human and property-based hardship. Six months after Katrina, New Orleans' population was 29 percent lower and stood at 181 400. This was slightly higher than the 136 681 residents who lived in the city four months after the storm.[15] The human toll extended far beyond simply the city limits, as Jefferson, Plaquemines and St Bernard Parishes all were severely impacted by the storm.[16] Population data for all ten parishes is listed in Table 14.1.

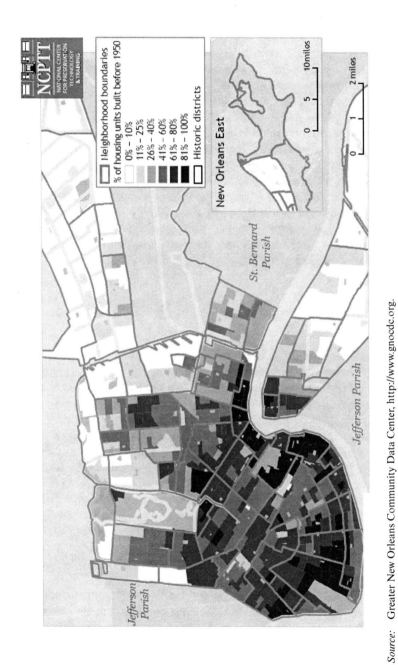

Source: Greater New Orleans Community Data Center, http://www.gnocdc.org.

Figure 14.2   Percentage of housing units built before 1950, by Census block group in Orleans Parish

Regarding housing, the Department of Housing and Urban Development (HUD) estimates that 71.5 percent of the 188 251 occupied housing units in Orleans Parish were damaged, including 78 810 that were severely damaged or destroyed.[17] Damage statistics for New Orleans, the three parishes that suffered severe population impacts and St Tammany Parish, which saw significant housing damage, are listed in Table 14.2.[18]

Within New Orleans proper, the level of damage sustained by neighborhoods can largely be measured by their ground elevations. Since there was not a significant amount of wind damage in New Orleans, neighborhoods located on the highest ground, encompassing the oldest areas of the city, were largely spared. Other low-lying neighborhoods, such as Lakeview and the Lower Ninth Ward, flooded because of their adjacency to failed canal levee sections, and nearby low-lying neighborhoods in the center of the city, such as Gentilly and Mid-City, were subsequently infiltrated by flood waters originating from the same sources. Storm surge along the Mississippi River Gulf Outlet Canal doomed New Orleans East, on the eastern side of the Industrial Canal.[19]

PROSPECTS FOR RECOVERY: ASSESSING THE DAMAGE

Given this background, this section seeks to describe the damage in ways that might guide decision-makers who are considering whether and how to reconstruct New Orleans. The approach here is to consider the city in a relatively disaggregated manner, looking at planning districts, which we treat as a proxy for neighborhoods, and within these districts, at the local geographic character. In particular, the analysis considers the extent to which damage within neighborhoods occurred in floodplains, which are clearly more likely to experience a repeat of the catastrophe inflicted by Katrina. The figures are based on a February 2006 review of conditions by the Department of Homeland Security in conjunction with the Federal Emergency Management Agency (FEMA), the Small Business Association (SBA) and the Department of Housing and Urban Development (HUD).[20]

Damage to Neighborhoods

We characterize the damage in a neighborhood by determining the fraction of units suffering damage and distinguishing between damage to owner-occupied and rental units. This distinction is important because of the varying sources of funding likely to be available in the reconstruction of each. Rental housing is frequently a corporate exercise involving large regional and

Table 14.1 Post-disaster population estimates for Greater New Orleans area

Parish	Feb 06	Jan 06	Dec 05	Nov 05	Oct 05	July 05	Change July 05–Feb 06	% change July 05–Feb 06
Jefferson	367 573	363 309	354 337	328 223	258 128	458 029	(90 456)	−19.75%
Orleans	181 400	156 140	138 681	138 681	138 681	458 393	(276 993)	−60.43%
Plaquemines	17 433	17 309	15 984	14 671	5 794	29 432	(11 999)	−40.77%
St Bernard	12 064	6 899	13 111	13 111	13 111	67 419	(55 355)	−82.11%
St Charles	51 517	51 830	52 137	52 205	52 492	48 359	3 158	6.53%
St James	21 659	21 747	21 981	21 975	22 317	20 842	817	3.92%
St John the Baptist	49 022	49 286	49 983	51 065	54 668	44 590	4 432	9.94%
St Tammany	206 204	205 461	202 536	198 668	180 513	213 633	(7 429)	−3.48%
Tangipahoa	108 674	109 423	110 359	110 547	111 982	103 232	5 442	5.27%
Washington	42 358	42 449	42 341	41 901	41 701	44 595	(2 237)	−5.02%
Total	1 057 905	1 023 854	1 001 451	971 047	879 388	1 488 524	(430 619)	−28.93%

Source: Compiled from the DHH/BPCRH Post Disaster Population Estimates, February 2006.

Table 14.2 Housing unit damage estimates for Greater New Orleans area

Parish	No. of occupied units	Minor damage	Major damage	Severe damage	Total damaged	% Occupied units w/ damage	% Occ. units w/ severe damage
Jefferson	176 234	59 552	29 643	4 677	93 872	53.3%	2.7%
Orleans	188 251	29 241	26 405	78 918	134 564	71.5%	41.9%
Plaquemines	9 021	2 033	1 190	3 994	7 217	80.0%	44.3%
St Bernard	25 123	561	5 938	13 748	20 247	80.6%	54.7%
St Tammany	69 253	31 182	15 948	1 682	48 812	70.5%	2.4%
Total	467 882	122 569	79 124	103 019	304 712	65.1%	22.0%

Source: Compiled from the US Department of Housing and Urban Development's Office of Policy Development and Research, February 2006.

national organizations that own portfolios of income-producing properties. As such, to the extent that New Orleans rebounds economically, parties interested in rebuilding rental units will likely have access to significant capital investment from abroad. By contrast, owner-occupied housing is much more disaggregated in terms of ownership and will require much more local capital. Given the human and economic toll borne by the area, such capital will likely be less forthcoming. Consequently, one might expect public subsidies to be required for this rebuilding at a level far greater than that for rental housing.

As a starting point, we define a neighborhood as suffering serious damage to a type of housing if at least 30 percent of that type of housing stock suffered some damage. By this definition, every neighborhood in New Orleans had serious damage to its owner-occupied housing stock (Table 14.3, column 3). Eleven of the city's 14 neighborhoods had serious damage to its rental housing stock, with only the Venetian Isles and the high-lying French Quarter and the Warehouse and Central Business districts not seeing this level of damage.

This sort of accounting might overstate the extent of damage, in the sense that the damage could be relatively minimal for many units in a given neighborhood. So next we consider neighborhoods as suffering extreme damage if at least 30 percent of their housing units suffered severe damage.[21] By this metric, we see that five neighborhoods no longer qualify as having extreme damage to their owner-occupied stock and six no longer had extreme damage to their rental housing stock. Even under this more restrictive definition, though, a majority of neighborhoods experienced extreme trauma and will face significant challenges in recovering.

Some might view the 30 percent threshold as somewhat arbitrary and perhaps not sufficiently high to identify those areas that experienced truly catastrophic damage. In recognition of this view, we establish a new threshold requiring that more than 50 percent of the housing stock experienced severe damage in order for viability to be a relevant issue. Neighborhoods facing this level of damage will be described as having catastrophic damage. A shown in the final two columns of Table 14.3, seven neighborhoods – Mid-City, Lakeview, Gentilly, Lower Ninth Ward, New Orleans East, Village de L'Est and Venetian Isles – had catastrophic damage to their owner-occupied housing stock and three had catastrophic damage to their rental stock. Three neighborhoods – Gentilly, Lower Ninth Ward and New Orleans East – had catastrophic damage to both types of housing.

The Floodplain versus Elsewhere

A review of the data suggests that, while a general aggregated neighborhood-level analysis is instructive, a further disaggregation might provide

Table 14.3 Damage to housing units, by type of housing tenure

	Number of damaged units		Damaged units as % of all units in district		Extreme damage?		Catastrophic damage?	
	Owner-occupied	Rental	Owner-occupied	Rental	Owner-occupied	Rental	Owner-occupied	Rental
Algiers	5 694	5 777	58.0	63.2	NO	NO	NO	NO
Bywater	4 784	6 715	74.6	74.6	YES	YES	NO	NO
French Quarter	255	461	35.7	21.1	NO	NO	NO	NO
Garden District	3 078	9 674	58.1	67.0	NO	YES	NO	NO
Gentilly	9 921	3 770	82.3	77.7	YES	YES	YES	YES
Lakeview	7 272	1 852	86.1	43.9	YES	YES	YES	NO
Lower Ninth	2 975	2 726	81.0	87.1	YES	YES	YES	YES
Mid-City	6 732	16 131	82.9	74.2	YES	YES	YES	NO
New Aurora	908	516	72.8	111.0	NO	NO	NO	NO
New Orleans East	13 621	12 730	86.3	97.3	YES	YES	YES	YES
Uptown	8 676	7 667	65.8	51.0	YES	NO	NO	NO
Venetian Isles	518	54	71.3	6.5	YES	NO	YES	NO
Village de L'Est	1 781	1 713	99.1	84.7	YES	YES	YES	NO
Warehouse/CBD	162	299	52.1	24.1	NO	NO	NO	NO
	(1)	(2)	(3)	(4)	(5)	(6)	(7)	(8)

Source: Data compiled by FEMA. SBA. HUD.

additional insights. In particular, given that much of the damage to the city
was due to flooding in low-lying areas, it might prove fruitful to consider the
level of damage to housing according to whether the housing was located in
a recognized floodplain or elsewhere in the city. This is done in Table 14.4.

As shown in the table, there is a striking difference in the extent of
damage to housing depending on whether the housing was located on a
floodplain. Only the Lower Ninth Ward and Village de L'Est neighborhoods had extreme damage to their housing stock in non-floodplain areas.
By contrast, nine neighborhoods saw extreme damage to their floodplain-
based owner-occupied housing stock and three had extreme damage to
their floodplain rental housing stock. Clearly, being located on a floodplain
was a key determinant of whether a neighborhood was likely to experience
extreme damage.

Turning to the question of viability, a similar picture emerges. No neighborhoods saw more than 50 percent of their non-floodplain housing suffer
severe damage, suggesting that viability for non-floodplain housing is not
a significant issue. On the other hand, the owner-occupied housing located
in floodplain areas of six neighborhoods suffered catastrophic damage.

*Table 14.4   Damage to housing units in floodplain, by type of housing
tenure*

	Extreme damage to housing in floodplain?		Catastrophic damage to housing in floodplain?	
	Owner-occupied	Rental	Owner-occupied	Rental
Algiers	NO	NO	NO	NO
Bywater	YES	NO	NO	NO
French Quarter	NO	NO	NO	NO
Garden District	NO	NO	NO	NO
Gentilly	YES	YES	YES	NO
Lakeview	YES	NO	YES	NO
Lower Ninth	YES	NO	NO	NO
Mid-City	YES	YES	YES	NO
New Aurora	NO	NO	NO	NO
New Orleans East	YES	YES	YES	YES
Uptown	YES	NO	NO	NO
Venetian Isles	YES	NO	YES	NO
Village de L'Est	YES	NO	YES	NO
Warehouse/CBD	NO	NO	NO	NO
	(1)	(2)	(3)	(4)

Source: Data compiled by FEMA, SBA, HUD.

Table 14.5   *Planning districts by percentage of total severe damage occurring on housing located in floodplains, by type of housing tenure*

| | % of all severe damage in floodplain | |
	Owner-occupied	Rental
Lakeview	95.1	94.7
Uptown	93.8	95.0
Garden District	92.7	32.8
Mid-City	92.7	93.4
Venetian Isles	92.0	73.8
New Orleans East	87.7	87.2
Gentilly	83.8	77.8
Bywater	69.5	70.0
Village de L'Est	66.0	30.8
Warehouse/CBD	50.0	76.0
Lower Ninth	45.4	39.9
New Aurora	22.2	18.8
Algiers	10.9	18.8
French Quarter	0	0
	(1)	(2)

Source: Data compiled by FEMA, SBA, HUD.

Interestingly, only in New Orleans East did the damage to the floodplain-located rental housing stock reach catastrophic levels.

Moreover, there is considerable variation among neighborhoods in terms of the degree to which damage was isolated to housing constructed on floodplains. More than 75 percent of all severe housing damage occurred in floodplain housing in Uptown, Mid-City, Lakeview, Gentilly, New Orleans East and Venetian Isles (Table 14.5). Significantly, this was true for both owner-occupied and rental housing in these districts. Damage in another district, Bywater, was similarly concentrated, with 70 percent of all severely damaged housing being located in floodplains.

By contrast, the French Quarter, Algiers and New Aurora neighborhood all had less than 25 percent of all the units experiencing severe damage located in floodplains. These neighborhoods tended to show less damage generally owing to their locations on relatively high ground, which also translated to their having relatively little floodplain within their boundaries.

The other planning districts fell somewhere between these two extremes. Alone among all districts, the Lower Ninth Ward had balanced damage between floodplains and other areas, with 40 percent and 45 percent of its severely damaged rental and owner-occupied housing, respectively, being located on floodplains. Differential floodplain impacts were found across owner-occupied and rental housing units in the Garden District (93 percent for owner-occupied versus 33 percent for rental), Village de L'Est (66 percent versus 31 percent) and the Warehouse and Central Business District neighborhoods (50 percent versus 76 percent).

What Should Be Rebuilt?

Given these data, the natural ensuing question is whether particular neighborhoods should be rebuilt and, if so, in what locations. The preceding analysis has highlighted several dimensions that must be considered. First, the breadth of damage to particular districts must be evaluated. If a majority of housing units in a neighborhood were destroyed, one must consider whether reconstructing an entire neighborhood is warranted given the economic and social realities that prevailed prior to the storm and those that currently exist. For example, it might not be desirable to rebuild a neighborhood that previously suffered from extremely high levels of poverty or unemployment, particularly if it is not apparent that opportunities and prospects will be substantially improved for those returning.

Second, the role of floodplain properties in determining the extent of damage is an important factor. Properties located on a floodplain will be the most susceptible to a repeat of the tremendous damage that was wrought by Hurricane Katrina. Consequently, one should give pause to notions of making significant investments in those floodplain communities that saw the most damage in this round. Moreover, if damage in a district were localized on floodplains, it might argue that the district might merit investment for rebuilding in all areas except the floodplain locales. Otherwise, the viability of the entire district might be called into question.

In helping policy-makers craft a solution to this key question, we outline how New Orleans' 14 planning districts fare along these dimensions and highlight those locations that have indicators suggesting rebuilding might not be optimal. Here, optimality is considered with an eye towards the scale of rebuilding required, minimizing the likelihood of another catastrophe, supporting the current economic and demographic foundations and overall cost. Costs are outlined in a separate section below. The goal is to provide an objective assessment of conditions.

Clearly there are other considerations that might come into play, such as the historical legacies of certain neighborhoods and communities. Class- and race-based equity issues in terms of the distribution of investments will certainly be important for those seeking to craft a solution. Economic incentives offered through federal and state sources might also make some

Table 14.6   Summary of degree of damage criticality, by planning district

	Extreme damage	Catastrophic damage	Extreme floodplain	Catastrophic floodplain	Floodplain concentration
Algiers					
Bywater	XX		X		X
French Quarter					
Garden District	X				X
Gentilly	XX	XX	XX	X	XX
Lakeview	XX	X	X	X	XX
Lower Ninth	XX	XX	X		
Mid-City	XX	X	XX	X	XX
New Aurora					
New Orleans East	XX	XX	XX	XX	XX
Uptown	X		X		XX
Venetian Isles	X	X	X	X	XX
Village de L'Est	XX	X	X	X	
Warehouse/CBD					

Source: Author calculations.

proposals more attractive than others. Though our exercise abstracts away from these issues, policy-makers will need to incorporate these and many other factors explicitly into any final decisions made regarding the disposition of neighborhoods and communities.

Table 14.6 summarizes how New Orleans's 14 planning districts fare along each of the dimensions introduced above. For each dimension, reflected as a column, 'X' indicates that the district exceeded the minimum threshold for being classified as in crisis, with 'XX' indicating a more critical condition. A neighborhood was considered to be in a more critical condition if both its owner-occupied and rental housing stocks exceeded the threshold for criticality.

The summary makes clear that some districts of the city essentially avoided significant damage.[22] These districts reflect the core of the old city (French Quarter, and Warehouse and Central Business District) and the area located south of the Mississippi River (Algiers and New Aurora). Any 'new' New Orleans should establish these districts as the primary core.

Slightly worse off than these core districts is the Garden District, whose rental housing stock experienced extreme damage and whose severely damaged owner-occupied housing was concentrated in floodplain areas. While this argues for some planning regarding the nature of rebuilding, this district still falls well below the damage experienced by other districts and thus likely should be an important portion of any rebuilding plan.

In assessing the status of the remaining nine districts, there are several approaches one might take. One could simply count the number of times that a district exceeded the minimum threshold for a dimension. By this measure, five districts met or exceeded the minimum threshold for criticality on all five dimensions and one district met criticality on four of the five dimensions. Except for the Mid-City district, these troubled districts all lie along Lake Pontchartrain. Mid-City itself was severely affected by flooding from the lake, suggesting that flood control and minimizing damage from flooding should be a primary consideration in the context of rebuilding.

An alternative approach might be to weight counts according to their severity, with higher severity along a dimension receiving more weight. If one were to use a simple scoring method assigning one point for each instance of criticality by housing tenure (that is, one point for criticality in owner-occupied housing and another point for criticality in rental housing), the results are largely the same. The same six planning districts identified as in crisis using the simple rule emerge as in crisis here. The only difference is that one additional district – the Lower Ninth Ward – also qualifies as in crisis, mainly a result of its suffering catastrophic damage to both its owner-occupied and rental housing stocks.

Bywater and Uptown, the remaining two planning districts, fall somewhat between the relatively minimally damaged core district and the ravaged lakeside districts. In both cases, the damage to the housing stock was heavily concentrated in floodplain areas. For example, about 95 percent of all housing damage in Uptown was to houses built in floodplain areas.

STRATEGIES FOR REBUILDING

An overarching strategy will be essential in order to develop, articulate and implement a rebuilding plan. This section describes possible strategies that could be pursued and provides estimates of cost, which will undoubtedly be a key factor in choosing between possible approaches. The goal is to lay out alternatives rather than select a preferred option. As the scope of this analysis is limited to housing, many other competing considerations that are outside of this narrow lens may influence the attractiveness or tractability of implementing particular approaches.

Initial Considerations

In considering whether and how to rebuild in New Orleans, one must consider a number of different factors. First, it is important to evaluate the likelihood that another Katrina-like disaster could occur. One consideration is the probability that other storms of Katrina's strength will form. Many experts believe that we are more likely to see category 5-level Gulf storms during annual hurricane seasons than have been typical historically.[23] This suggests that the risk of another storm of Katrina's power hitting New Orleans is greater than in past years.

The viability of any rebuilding plan will rely critically on the degree to which rebuilt structures will be protected in the event of another storm of Katrina's magnitude. Because most damage resulted from a failure of the city's flood mitigation mechanisms, a key issue in this regard is levee reconstruction. New Orleans' levees were designed to withstand powerful storms, but age, limited maintenance and design flaws reduced their ability to protect the city when such a large storm hit. If the levees are rebuilt to the same standards and managed in the same way, housing units will remain as exposed to catastrophe as they were pre-Katrina. More sturdy levees will clearly reduce risk to housing, but will be considerably more expensive to build. This trade-off will have to be weighed by policy-makers.

A third issue is the extent to which economic activity can adequately support housing demand at prevailing market prices. Many New Orleans neighborhoods lacked considerable vitality prior to Katrina, with some areas having unemployment rates five or six times the national average, and many of these were precisely those that saw the most extensive damage.[24] The prospects for improved economic performance in these areas are undoubtedly worse than prior to the storm. Moreover, other neighborhoods whose employment and economic vitality relied on broader regional economic activity might be jeopardized if business and industries do not return in pre-storm numbers. This issue is highlighted in a recent Brookings Institution report, which notes that the New Orleans labor force has contracted by 30 percent and is only growing at a relatively slow pace.[25] Thus, there is considerable uncertainty regarding the level of housing investment warranted given the city's economic realities.

Related to this is the fact that land and construction costs will establish a minimum price for housing. These costs have rapidly risen since the hurricane, due in part to supply shortages driven by pressures from abroad. Housing is therefore more costly to produce, purchase and rent, which requires city residents to have sufficiently high incomes to afford them. Without either construction or financing subsidies, areas with weakened economies are less able to support such housing.

Building quality is yet another issue to be considered, as homes built to withstand extreme winds and avoid flooding are less likely to sustain damage. Here, steps are being taken to minimize existing risks. The FEMA has recommended that all new construction and all substantially damaged structures be elevated to either the base flood elevation as currently stipulated or at least 3 ft above the highest adjacent existing ground elevation at the building site, whichever is higher.[26] In addition, in the wake of the hurricane, in an attempt to secure federal funding and lure insurance companies back into Louisiana, the state adopted the International Building Code (IBC) in November 2005. This code applies to all new homes in New Orleans as well as those that sustained more than 50 percent wind damage.

These code changes, while reducing the likelihood of structural damage, will result in higher redevelopment costs. Estimates for raising existing homes to meet the FEMA advisory standards vary from $8000 to $100 000 depending on location, home size and foundation type.[27] The Home Builders Association of Greater New Orleans estimates that raising new homes 3 ft above the ground will increase development costs by 10 to 15 percent.[28] These elevated costs are in addition to those associated with meeting the new IBC.

There is some cost mitigation available to those interested in building or rebuilding in the city. The FEMA offers assistance of up to $30 000 to raise existing structures to code, low-interest Small Business Administration (SBA) loans are also available through the FEMA for amounts up to $200 000, and Community Development Block Grant funds are likely to be available for rebuilding assistance.[29]

Strategic Options

We begin our assessment of the various options for redevelopment by making a number of assumptions. First, it seems reasonable to assume that the likelihood of another storm of Katrina's magnitude is sufficiently high and that the new building codes for any new housing construction and for severely damaged homes will be enforced. For the initial assessments, we will further assume that levees will be rebuilt to the pre-storm standards. We will revisit this issue subsequently.

From an economic perspective, given the tremendous economic disruption, it would seem unlikely that the post-disaster level of activity will reach pre-storm levels. This suggests that fewer housing units will be needed than were originally in place in the city. Of course, the outstanding and critical issue is how many fewer. For simplicity, we will consider cases where the city requires 75 percent, 85 percent or 95 percent of the number of pre-storm housing units. As there were approximately 188 000 housing units in

Table 14.7   Units produced under various reconstruction strategies

Planning District Units to redevelop	Non-devastated areas	Non-devastated areas plus Mid-City district
All units	60 367	83 230
All units not severely damaged	35 870	43 165
All non-floodplain units	37 584	40 724
All non-floodplain units not severely damaged	29 114	31 196

Source: Author calculations.

Note: Non-devastated areas include the French Quarter, Warehouse/Central Business District, Garden, Uptown, Bywater, Lower Ninth, Algiers, and New Aurora Planning Districts.

New Orleans before Katrina, this implies that the city will need 141 000, 160 000 or 179 000 units post-reconstruction. The FEMA assessment indicated that there were 54 000 undamaged homes after the storm, which means that reconstruction efforts must produce 87 000, 106 000 or 125 000 units in order to meet the city's needs, depending on the economic scenario one uses.

If the levees are built to pre-existing standards, then housing units in the planning districts that the prior analysis determined to exceed criticality along the most dimensions, if rebuilt, would be at considerable risk. Thus, one potential approach would be to avoid reconstruction investment in these areas altogether and limit investments to those neighborhoods and districts that remained somewhat viable in the wake of the storm, which we will designate as 'non-devastated areas' (NDAs). The Bywater and Uptown districts would be included as NDAs, under the reasoning that the improved building standards would reduce likely impacts in these areas considerably.

Within NDAs, an issue is what exactly to rehabilitate. One could choose to rehabilitate and rebuild everything in these areas. However, there is considerable variation in the cost of reconstruction, with much higher costs for severely damaged structures than for other structures. Thus, one might opt to focus investments on all units that were not severely damaged. Finally, one must decide whether to promote either type of construction in floodplain areas.

Table 14.7 shows the number of units that would be produced under these four scenarios. None of these scenarios produces sufficient housing units to meet the economic need. Thus, redevelopment in other parts of the city will also be necessary. Given the city's geography, Mid-City would be a

Table 14.8 Units produced under reconstruction strategies based on damage

Units to rehabilitate	Number of units produced
Minor damage only	28 958
Minor or major damage	54 823
Minor or major damage, severe damage if in NDA	79 320
Minor or major damage, severe damage if in NDA or Mid-City district	94 888

Source: Author calculations.

Note: NDAs are non-devastated areas and include the French Quarter, Warehouse/ Central Business District, Garden, Uptown, Bywater, Lower Ninth, Algiers, and New Aurora Planning Districts.

natural district to add to the NDAs. From Table 14.7, it is clear that we begin to approach the level of units needed if Mid-City is treated as an NDA, but still fall well short of more optimistic economic scenarios.

A completely different and more cost-based approach would be to prioritize reconstruction according to the extensiveness of needed repairs. Under this approach, the focus would first be on units suffering minor damage, then on those with major damage, and lastly on those with severe damage. The results of this are shown in Table 14.8. Clearly, limiting attention to only non-severely damaged units does not yield adequate housing for the city. If one were to be conservative and add units severely damaged located in NDAs with or without Mid-City, one begins to approach the desired number of units. One difficulty with this latter approach is that it is less attuned to geography and could leave some communities particularly isolated. Thus, care would be required in implementing this type of approach.

As a final approach, if one combines the two approaches, rehabilitating all units in NDAs and Mid-City plus housing units in other districts that suffered minor or major (that is, non-severe) damage, the city would have 102 183 available units. This is close to meeting the housing needs under the moderate economic forecast.

It should be obvious at this point that producing the required number of units under the more optimistic economic scenarios will be difficult under the initial assumptions. The needs of New Orleans can be met, however, if one were to alter assumptions regarding levee reconstruction and building patterns. If the levees are built to a higher standard, then rebuilding outside of the NDAs becomes less risky and larger-scale investment in housing in these areas becomes more tenable. Exactly how much more tenable will depend on the nature of the improvements. A second consideration is the pattern of redevelopment. The scenarios above have implicitly assumed that reconstruction would occur in the densities that prevailed prior to the storm. If, however, development was more dense in the NDAs and other areas, making New Orleans a more compact city, then the desired level of housing could certainly be accommodated. There is an important role that zoning, parking regulations and other land use planning issues will play in determining the likely success of any of these strategies in adequately serving New Orleans.

Costs of Redevelopment

The previous section presented estimates of the number of units that would be produced under various district-based and damage-based decision rules. One must consider the cost of each strategy in order to compare them properly. This requires obtaining estimates of rehabilitation costs for homes with minor and major damage and replacement costs for homes that suffered severe damage. In the course of assessing damage and claims for homes in New Orleans, the Small Business Administration hired inspectors to assess property loss for a sample of properties. Cross-listing these assessments with the FEMA's categorization of the nature of a home's damage, one can estimate the cost of redevelopment. Based on these data, rehabilitation costs should be approximately $84 000 per housing unit and reconstruction costs will be about $100 000 per unit.[30]

These figures require some adjustment. First, data from other sources suggests that the reconstruction cost estimate is likely low. For example, a recent comparative study of construction costs found the cost for new townhouse construction is almost $200 000 for Dallas, a market that is arguably comparable to New Orleans in terms of land values and labor and materials costs.[31] This divergence is undoubtedly a result of SBA inspectors considering rehabilitation rather than demolition and new construction. Second, costs will also vary depending on the density of development, with lower-density development being more expensive on a per-unit basis. Recent estimates suggest a multiple of about 40 percent.[32] However, if we assume that the distribution of housing by tenure in the sample matches the distribution of units to be produced in the scenarios, then no adjustment on this dimension is needed.[33]

Without a clear rule for establishing new construction costs, we take as a benchmark $150 000. This strikes a balance between the SBA inspector estimate and the market-based assessment, and reflects a balance between rehabilitation and reconstruction in how housing investments are structured. We note that this figure matches the maximum amount of post-insurance

Table 14.9   Estimated costs of redevelopment under various strategies (in $ billions)

Redevelopment strategy	Baseline	Lower rehab costs
District-based		
NDA only		
All	6.69	5.18
All non-severe	3.01	1.51
All non-floodplain	3.72	2.49
All non-floodplain, non-severe	2.45	1.22
NDA plus Mid-City		
All	9.64	7.82
All non-severe	3.63	1.81
All non-floodplain	4.05	2.74
All non-floodplain, non-severe	2.62	1.31
Damage-based		
Minor	2.43	1.22
Minor and major	4.61	2.30
Minor and major plus NDA severe	8.28	5.98
Minor and major plus NDA severe plus Mid-City severe	10.61	8.31

Source: Author calculations.

Note: NDAs are non-devastated areas and include the French Quarter, Warehouse/ Central Business District, Garden, Uptown, Bywater, Lower Ninth, Algiers, and New Aurora Planning Districts. All estimates assume a reconstruction cost of $150 000 per unit. The baseline estimates assume a rehabilitation cost of $84 000 while the lower rehab cost is set at $42 000.

reconstruction money available to households from the state of Louisiana via its recently established The Road Home Program.[34] Thus, the benchmark value used here comports with public policy assessments to some degree.

We will use the SBA's $84 000 estimate as an average cost for all housing units suffering minor and major damage. Because the SBA estimates were for units with major damage, we also produce estimates using a rehabilitation cost of $42 000 on the assumption that homes with minor damage might require considerably less work.

The results of these estimates, shown in Table 14.9, suggest that the cost of reconstructing New Orleans will be substantial under any scenario. Even the least-costly scenario of only fixing housing units with minor damage will require $1.2 billion. Depending on the estimate of cost for rehabilitation, the scenario that gets closest to meeting New Orleans' need for housing based on economic considerations will cost between $8.3 billion and $10.6 billion.

These estimates are likely a lower bound for costs. The FEMA assessments of damage were frequently based on the value of the damaged property rather than the cost of rebuilding the structure. Because of this, properties with low market values might not meet the threshold of severe damage ($30 000 in property damage) even if rebuilding might require higher outlays. This is a particular issue in the Lower Ninth Ward, in which about 36 percent of housing units were valued at below $50 000 in 2000 according to the Census. Unless these structures were a total loss, it is reasonable to assume that some that suffered significant damage and will require close to complete reconstruction did not meet the $30 000 threshold. In these cases, the cost of rebuilding will be higher than estimated here.

While public subsidies will certainly not fully provide the estimated $8.3 billion to $10.6 billion needed to approach New Orleans' projected housing needs based on economic considerations, given the hardship that many households have undergone and the declining availability of insurance for property owners, substantial public involvement will be needed. The public sector has already provided considerable resources. For example, the Road Home Program is a $10.4 billion commitment of federal funds for Louisiana recovery. However, note that this allocation is for the entirety of Louisiana, whereas our estimates suggest that an equal amount will be required for New Orleans alone. Thus, the magnitude of investment is indeed massive.

CONCLUDING THOUGHTS

The devastation of New Orleans' housing stock caused by Hurricane Katrina was of an unprecedented scale, and reconstruction will likewise require a massive effort and commitment along human, financial and public sector dimensions. The public sector must facilitate the creation of a vision regarding the character and physical form of the 'new' New Orleans and establish incentives to support this vision. Individual households must embrace this vision and commit their resources to rebuilding in concert with it. And financial institutions, including banks and insurance companies, must also provide key backing to those seeking to implement the vision and redevelop the city.

The role of the insurance sector is particularly critical. Because at this time major companies are generally not writing new homeowners policies, most buyers will pay a premium (two to three times as much) for insurance through Louisiana Citizens Property Insurance Corp, which raises the cost of ownership.[35] The staggering cost of insurance could thus be an important impediment to rebuilding efforts.

The present analysis of the housing sector has established a framework that can contribute to the discussion shaping the vision of the 'new' New Orleans. Various strategic approaches have been introduced, different scenarios were considered, and their costs compared. These can be instructive for decision-makers that must juggle the many different interests of their constituents.

A key component in this process will be determining the final post-Katrina economic capacity of New Orleans. The analysis here included three scenarios of the city's capacity, and the extent of the challenge in producing the needed volume of units varied across them. Policy-makers will need to devote considerable resources to grasp the nature of the economic response to New Orleans' hardships.

In closing, we raise a few additional important issues. First, housing affordability will be a considerable concern moving forward. In the 'high-and-dry' areas of the city, the New Orleans real estate market is booming. For example, home prices in the Garden and Uptown Districts had a 49 percent year-over-year increase in the last quarter of 2005.[36] As a result, New Orleans is facing a housing crisis, particularly for low-wage workers. While there is significant demand for housing in the $100 000 to $200 000 range, there is limited product available.[37] Moreover, the vacancy rate for rental units is virtually zero, compared to a 7 percent vacancy rate before the storm.[38] The US Department of Housing and Urban Development (HUD) has increased the fair market rents and the rent subsidy for the eight-parish New Orleans MSA by approximately 35 percent post-Katrina.[39] However, critics argue that these will do little to solve the housing problem because there simply are not enough units to meet the demand.[40] Before Katrina, about 9000 families in New Orleans relied on Housing Choice Vouchers (Section 8); 4300 vouchers have been issued since, but only 1600 have been used.[41] Thus, living in New Orleans has become significantly more difficult for those with the lowest means.

Closely related to this is the question of who occupies the rebuilt housing. The housing voucher experience suggests that housing provided at market rates would be beyond the means of many pre-Katrina New Orleans residents. Such an outcome would leave many of the poor (and black) residents bearing a disproportionate burden in terms of life disruption caused by the storm. Policy-makers will have to consider carefully how to craft a rebuilding strategy that maintains some degree of social and economic equity while meeting market standards to ensure that redevelopment occurs.

Second, more attention should be placed on incorporating 'passive survivability' into new and existing structures. Promoted by the US Green

Building Council, passive survivability ensures that a building can maintain critical life-support conditions for its occupants if services such as power, heating fuel or water are lost for an extended period.[42] Many mechanisms for implementing passive survivability, such as having significant natural ventilation, a highly efficient thermal envelope, passive solar gain and natural daylighting, are largely not incorporated into current building codes, but could do much to reduce the impact of disasters and terrorist attacks, particularly to households.[43] While relevant for homes, this quality is especially important for buildings identified as emergency shelters. For instance, temperatures reached 105°F in the Superdome where thousands sought shelter during the storm and in the days following.[44]

Third, the analysis here suggests that New Orleans' West Bank – the Algiers and New Aurora Districts – should be a central part of any reconstruction strategy for the city, as it was spared much of the devastation. It is important to note that the West Bank also has flood zones and relies on levees for flood control. If the storm had taken a different path, it is very possible that this area might have seen comparable damage and disruption. Planners should consider this very carefully, including a study of the existing levees and the likely paths of storm surges, before crafting a final strategy for reconstruction and rebuilding.

Finally, as a general issue, this catastrophe makes clear the need for better and more detailed disaster planning by cities and regions across the nation. Proactive risk assessment, evacuation planning and intergovernmental coordination can go a long way toward minimizing the likelihood of another urban disaster at the scale of Hurricane Katrina in New Orleans.

NOTES

1. Knabb, Richard D., Jamie R. Rhome, and Daniel P. Brown (2005), 'Tropical cyclone report: Hurricane Katrina', National Hurricane Center, 20 December, pp. 6–9.
2. 'The federal response to Hurricane Katrina: lessons learned', February, 2006, p. 1.
3. Gabe, Thomas, Gene Falk and Maggie McCarty (2005), 'Hurricane Katrina: social-demographic characteristics of impacted areas', Congressional Research Service, 4 November, retrieved 11 February 2006 from http://www.gnocdc.org/reports/crsrept.pdf.
4. Ibid.
5. Ibid.
6. Knabb, Richard D., Jamie R. Rhome and Daniel P. Brown (2005), 'Tropical cyclone report: Hurricane Katrina', National Hurricane Center, 20 December, pp. 6–9.
7. 'Historic preservation vs. Hurricane Katrina', (2005), *Congressional Quarterly*, Retrieved 16 February 2006 from LexisNexis Academic.
8. Knabb, Richard D., Jamie R. Rhome and Daniel P. Brown (2005), 'Tropical cyclone report: Hurricane Katrina', National Hurricane Center, 20 December, pp. 6–9.
9. US Department of Housing and Urban Development (2006), 'Current housing unit damage estimates', 12 February, retrieved 25 March 2006 from http://www.gnocdc.org/reports/GulfCoast_HousingDamageEstimates_021206.pdf.

10. Louisiana Department of Health and Hospitals Bureau of Primary Care and Rural Health (2006), 'Post disaster population estimates', February, retrieved 25 March 2006 from http://www.gnocdc.org.
11. Preservation Resource Center, Architecture, retrieved 23 April 2006 from http://prcno.org/arch.html.
12. Ibid.
13. Lewis, Pierce (2003), *New Orleans: The Making of an Urban Landscape*, Santa Fe, NM: Center for American Places.
14. US Census.
15. Ibid.
16. Of these, Jefferson Parish is a somewhat special case. Despite not having any major levee breaches, the area flooded because of a mayoral decision to evacuate pump stations in advance of the storm.
17. US Department of Housing and Urban Development (2006), 'Current housing unit damage estimates', 12 February, retrieved 25 March 2006 from http://www.gnocdc.org/reports/GulfCoast_HousingDamageEstimates_021206.pdf.
18. The extensive damage in non-Orleans parishes reinforces the notion that maintaining a broader regional perspective is important. While 42 percent of Orleans Parish housing suffered at least severe damage, an even greater percentage of housing units were destroyed in the neighboring parishes of St Bernard and Plaquemines. In addition, any plan for the city's rebuilding must also consider Jefferson Parish in the process since it is currently the most significant portion of the Greater New Orleans area population and home to many of Orleans Parish's workers.
19. This storm surge also devastated St Bernard Parish.
20. 'Current housing unit damage estimates: Hurricanes Katrina, Rita, and Wilma', US Department of Housing and Urban Development, 7 April, 2006 revision.
21. A housing unit is considered to have suffered severe damage if property damage exceeded $30 000 or, if there were no direct inspection, if remote sensing indicated a water depth of at least two feet.
22. It is important to remember, however, that all districts saw at least 30 percent of its housing stock suffer some damage. The distinction for these four districts is that the vast majority of the damage was of a relatively minor sort, either requiring less than $5200 of expense per housing unit or involving a unit with flooding of less than 1 ft.
23. For example, see Kerry Emanuel (2005), 'Increasing destructiveness of tropical cyclones over the past 30 years', *Nature*, **436** (7051), 686–8.
24. Greater New Orleans Community Data Center, compiled from US Census Bureau Census 2000 SF3 data sample.
25. Amy Liu, Matt Fellows and Mia Mabanta (2006), 'Special edition of the Katrina Index: a one-year review of key indicators of recovery in post-storm New Orleans', August, Metropolitan Policy Program report, Brookings Institution.
26. FEMA (2006),'FEMA flood recovery guidance', 12 April, retrieved 23 April 2006 from http://www.fema.gov/pdf/hazard/flood/recoverydata/orleans_parish04-12-06.pdf.
27. Thomas, Greg (2006), 'Home builders decry 3-foot elevation rule', *Times-Picayune*, 22 April, retrieved 22 April 2006 from http://www.nola.com.
28. Theyenot, Brian (2006), 'Finally, rules for rebuilding', *Times-Picayune*, 13 April, retrieved 13 April 2006 from http://www.nola.com.
29. Ibid.
30. These figures were calculated by calculating a weighted average of the median cost estimates for each of the New Orleans planning districts as appropriate.
31. Jerry J. Salama, Michael H. Schill and Jonathan D. Springer (1991), 'Reducing the cost of new housing construction in New York City', Furman Center for Real Estate and Urban Policy report, New York University.
32. Ibid.
33. The scenarios in the previous section suggest an ownership rate of between 40 percent and 47 percent.

34. The Road Home Program sets aside a fraction of the state's Community Development Block Grant program allocation for Louisiana owners of homes destroyed or damaged by Hurricanes Katrina and Rita. Up to $150 000 is available to eligible households. The program also sets aside money to support rental housing, the homeless, and developers seeking to work in distressed Louisiana neighborhoods. The program is not limited to New Orleans. HUD News Release (2006), 'Jackson approves Louisiana's $4.6 billion Road Home Program', 30 May, http://www.hud.gov/news/release.cfm?content=pr06-058.cfm
35. Mowbray, Rebecca (2006), 'At a premium', *Times-Picayune*, 29 March, retrieved 2 July 2006 from http://www.nola.com.
36. Thomas, Greg (2006), 'New Orleans real estate surged after Katrina', *Times-Picayune*, 21 February, retrieved 2 July 2006 from http://www.chron.com/disp/story.mpl/business/3672824.html.
37. Thomas, Greg (2005), 'Real estate market gets hotter after storm passes', *Times-Picayune*, 17 December, retrieved 2 July 2006 from http://www.nola.com.
38. Stone, Richard (2006), 'New Orleans real estate market overview', NAI/Latter and Blum's Market Snapshots, 1 June retrieved 2 July 2006 from http://www.latterblum.com/markets/no.html.
39. Greater New Orleans Community Data Center (2006), 'Metro New Orleans fair market rent history', retrieved 2 July 2006 from http://www.gnocdc.org/reports/fair_market_rents.html. Filosa, Gwen (2006), 'HUD raises rental voucher values', *Times-Picayune*, 22 June, retrieved 2 July 2006 from http://www.nola.com.
40. Ibid.
41. Ibid.
42. Wilson, Alex (2005), 'Passive survivablity', *Environmental Building News*, December, retrieved 21 January 2006 from http://www.BuildingGreen.com.
43. Ibid.
44. Ibid.

15. Unnatural disaster: social impacts and policy choices after Katrina

John R. Logan

Early media reports about the wind damage and flooding caused by Hurricane Katrina focused on New Orleans, and especially on the people who had been unable to escape the city before it flooded. Images of poor and predominantly black people crowded into the Superdome and Convention Center created indelible impressions about who was affected most strongly. We now know that most residents had evacuated safely and that even some mostly white and predominantly middle-class neighborhoods were decimated by the flooding. And yet the initial impression is true: Katrina disproportionately affected poor, black neighborhoods. Because these are the residents with the least market resources, this means that policy choices affecting who can return, to which neighborhoods, and with what forms of public and private assistance, will greatly affect the future character of the city.

What stands out is the failure to formulate a coherent policy. But what is visible so far is disturbing. It now appears that the recovery of New Orleans will be unusually slow. A reliable estimate of the city's population prepared by the Louisiana Department of Health and Hospitals (2006) estimated a total of 201 000 persons, far less than half the 494 000 counted by Census 2000. The white population in the period June–October 2006 was about two-thirds of its former size, while the black population was down by nearly three-quarters. Several questions will be addressed here to understand this result and interpret its future implications: (1) Who was displaced by Katrina? (2) How does the pattern of displacement affect people's chances of returning? (3) How are public policy decisions affecting the recovery process? and (4) What do shifts in local political influence portend for the future?

DISPLACEMENT FROM NEW ORLEANS

The best information about displacement, including the neighborhood location and social composition of affected people, is based on counts of the population living in areas that were flooded. My estimate is that nearly 650 000 persons lived in heavily damaged areas in the New Orleans metropolitan region and Mississippi coast (Logan, 2006a). More than half of these, 354 000, lived in the city proper, Orleans Parish (for a similar conclusion, see the report by the Congressional Research Service: Gabe et al., 2005).

Figure 15.1 shows flooded and non-flooded areas of Orleans Parish along with the racial composition (percent black) of census tracts. This map shows that the undamaged areas of the city were mainly in two areas. One is just north of the Mississippi River in a zone extending westward from downtown. The other is across the river on the West Bank, in a district known as Algiers. The map shows that some predominantly white neighborhoods in the northwest part of the city were entirely flooded. However almost all of the neighborhoods that were in the range of 75 percent to 100 percent black at the time of Census 2000 were flooded. I estimate that about 265 000 of the city's Census 2000 black population of 325 000 lived in flooded zones. This compares to about 68 000 of 129 000 non-blacks.

Separate analyses demonstrate that damaged areas were also disproportionately composed of renters and lower-income residents. However it is the division by race that stands out most strongly, because the most damaged black neighborhoods had varying class composition, ranging from predominantly middle-class New Orleans East to the much less affluent Lower Ninth Ward, to neighborhoods with public housing projects where a majority of residents were below the poverty line.

Discussions of the racially differential impact of Katrina have often emphasized the Lower Ninth Ward (where many homes were entirely demolished by the breach in the levee of the Industrial Canal) and New Orleans East. Most neighborhoods in these planning districts were more than 85 percent black, and most residences were damaged. A majority of residents of both of these planning districts were homeowners, though there were clear class distinctions between the two areas. More than a third of Lower Ninth Ward residents were below the poverty line, and nearly 14 percent were unemployed. New Orleans East had a considerably larger middle-class component, though it was not among the city's most affluent sections.

Many of the most segregated neighborhoods with the highest poverty rates are those identified as 'projects', a reference to the prominence of public housing within their borders. The project neighborhoods typically had poverty rates in the range of 60–80 percent of the population, unemployment is above 20 percent, they were all predominantly black (with African Americans accounting for 90 percent or more of their residents),

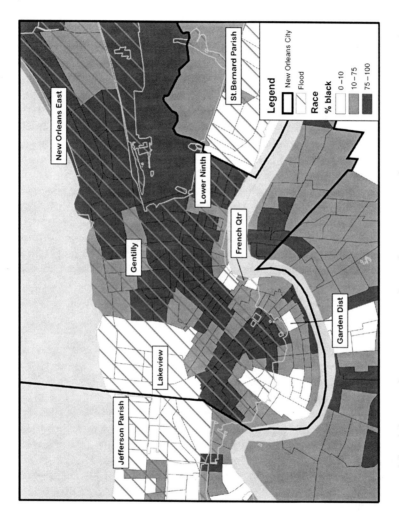

Figure 15.1 *Extent of flood damage from Katrina and racial composition of census tracts in New Orleans*

Legend

New Orleans City

Flood

Race
% black

0–10

10–75

75–100

New Orleans East

St. Bernard Parish

Lower Ninth

Gentilly

French Qtr

Garden Dist

Lakeview

Jefferson Parish

281

and 80 percent or more of residents were renters. There are six such neighborhoods in New Orleans (though there are concentrations of public housing or Section 8 housing [i.e. Federal government's largest rental subsidy program] in other parts of the city). In five of them with a combined 2000 population of over 15 000 persons (Calliope, Iberville, St Bernard, Desire and Florida) the entire territory was damaged.

At the other end of the class spectrum are a number of more advantaged neighborhoods with poverty rates below 10 percent or unemployment rates below 5 percent. In the most heavily impacted planning districts, only a few neighborhoods meet either criterion. These include the Lake Terrace and Lake Oaks neighborhood in Gentilly and the Read Boulevard East neighborhood in New Orleans East. Most such neighborhoods are in the Lakeview Planning District, which is an area with a small black population, mostly homeowners, and very low rates of poverty and unemployment. Here only the Lakeshore and Lake Vista neighborhood, adjacent to Lake Pontchartrain, was partly spared.

Few residents in the French Quarter, a predominantly white neighborhood with a poverty rate of about 11 percent and unemployment below 5 percent, lived in tracts that were flooded. Among other neighborhoods with a national reputation for affluence, the Garden District neighborhood was not flooded and only 40 percent of the Audubon and University neighborhood (home of Tulane University and Loyola University) was damaged.

WHERE PEOPLE WENT

The census data represent the numbers of persons who were at greatest risk of being displaced for more than a few weeks. What is known about the actual long-term displacement of population? Evidence is given here from three different sources. Each source has its own limitations, but taken together these sources offer a consistent picture: the majority of the city's population is still living elsewhere, of these the largest share is living outside the state, and black residents (especially poor black residents) are disproportionately found at the greatest distance from their prior homes.

One source is postal change of address data in the post-Katrina period, tabulated by the US Postal Service (Russell, 2006a). These data identify the original pre-Katrina three-digit zip code (origin) and current three-digit zip code (destination) of households that filed changes of address. At the end of March 2006 more than 160 000 households were relocated from their original address in Orleans Parish. Of these, about 17 000 were at a new address within Orleans Parish. About 21 000 were elsewhere in the metropolitan region, plus 15 000 in Baton Rouge, and 12 000 in other parts of

Louisiana. Close to two-thirds were out-of-state, most prominently in Texas (52 000). The most common out-of-state destinations were Houston (27 000), Dallas (14 000) and Atlanta (8000).

Another source is the Federal Emergency Management Agency (FEMA)'s tally of the reported addresses of area residents who had applied for assistance. This information was prepared in mid-February 2006 and made available in the federal court case that challenged election procedures (*Wallace* v. *Blanco*). An astounding total of nearly 400 000 persons initially living in Orleans Parish had applied for assistance. Of these, 154 000 were living within Louisiana, including a number of persons who had suffered relatively minor damage and returned to their original homes. But over 100 000 reported addresses in Texas and an even larger number were living in other states. These numbers reinforce the conclusion above about the significance of displacement outside of Louisiana, especially to Texas.

The impacts of displacement depend not only on its volume but also its location – and the furthest away turn out to be African Americans, especially those with the lowest incomes. The only public source of information about the racial composition and income levels of displaced persons is the Current Population Survey (CPS), conducted by the US Department of Commerce. The CPS is collected monthly for a national sample of 60 000 households. It is designed to be representative of the civilian non-institutional population aged 16 and above. Beginning in November 2005 the CPS included a question to identify persons who were evacuated as a result of Hurricanes Katrina and Rita. Its principal limitations are its relatively small sample size and its exclusion of persons living in shelters, hotels or other forms of group quarters. The sample weights provided by the Bureau of Labor Statistics allow the sample to be used to produce population estimates, and I have used the sample data from December 2005 to evaluate the racial composition and income levels of displaced persons (Logan, 2006b). I selected only persons whose original pre-Katrina residence was in the state of Louisiana. I focused on non-Hispanic whites and non-Hispanic blacks. The number of evacuees identified as Hispanic, Asian or other race is too small to permit analysis.

In December 2005, evacuees identified in CPS-sampled households represented about 1.1 million persons aged 16 and over who had evacuated from where they were living in August. Just over half of these persons had returned to the home from which they had evacuated. According to this source, more than 400 000 whites were evacuated, of whom 67 percent had returned home. Among blacks, more than 200 000 were evacuated, of whom less than 40 percent had returned home. Of those who were still displaced, nearly two-thirds of whites were in Louisiana, while three-quarters of blacks were out-of-state.

It is also relevant to compare the income levels of white and black evacuees. White evacuees had similar income levels regardless of their current location, with a median just under $50 000 and less than 15 percent in households with income under $20 000. Black evacuees who had returned home had somewhat lower incomes than whites. Blacks who remained displaced had much lower incomes, with a median of under $15 000 and more than 60 percent below $20 000. It is clear, therefore, that blacks – because they were further away and had fewer personal resources – faced great obstacles to their return to New Orleans.

POLICY CHOICES

The sheer number of people who lived in heavily damaged areas is a reminder of the scale of Katrina's impact. Because the storm hit large numbers of people of every race and class, it is not surprising that initial public support for policies to assist these people also cut across race and class lines. However there was also a substantial disproportionate impact on African Americans and people with fewer resources. These disparities stem from within the city of New Orleans itself, and more specifically from vulnerability to flooding. This is a pattern with deep roots, and although Katrina caused the most extensive flooding in memory, prior studies by historians (such as Colten, 2005) have demonstrated that both high ground and public investments in drainage and pumping systems consistently worked to the advantage of certain neighborhoods in past storms.

There are major variations across the region that are likely to affect the process of recovery. Damage was extensive on the Mississippi coast, and the area's largest single source of employment – casino gambling – was knocked out of operation. In comparison to New Orleans, however, the number of people living in areas of moderate or greater damage was small, only about 50 000. And also in contrast to New Orleans, only a small share of these people were black and a majority were homeowners. It is difficult to assess the importance of race in recovery policy in Mississippi, but in a politically conservative state it could make a big difference that white homeowners constitute the bulk of claimants for state assistance. Further, these people are easier to serve for several reasons:

1. First, they are identifiable and – because they retain an ownership interest in their properties – they should prove easier for authorities to contact.
2. Second, since much of the damage wrought by Katrina in this area was by wind and rain damage, standard homeowner policies offer

substantial private sector coverage of damage losses. For those with uninsured flood damage, Mississippi state government expects federal aid to
be sufficient to fund payments of $150 000 to individual homeowners.
3. Third, the low density of housing in this area means that typically even
    when one's home was uninhabitable, there was space for a trailer in the
    driveway. Since in addition the loss of electrical power was relatively
    short term in Mississippi, and basic public services could be restored
    within a reasonable time, homeowners in this region more readily met
    the requirements for a FEMA-provided trailer – space and confirmed
    utility hookups.

In contrast, consider the situation in New Orleans. Six months after
Katrina, observers were beginning to see signs of progress (Russell, 2006b).
But more than half the persons in damaged areas were renters, unlikely to
be protected in any way by property insurance, and 30 percent fell below
the poverty line and were therefore unlikely to have their own funds to
return to the city. By the end of 2005, power was still unavailable much of
the city, and actual connections to electric power required residents to
present evidence of inspection by a licensed electrician before power would
be restored to an individual home. The utility company (a subsidiary of the
Entergy Corporation) had filed for bankruptcy protection in September.
Large areas of the city remained vacant even at the beginning of 2007.
Though most debris had been removed and many homes had been gutted,
reconstruction work had not begun on most homes. Basic public services
had not been restored in many neighborhoods. For example, only 54 public
schools reopened in autumn 2006 (compared to 128 before Katrina). As
shown in Figure 15.2, these schools were concentrated in high-ground
neighborhoods close to the Mississippi River, with almost none in New
Orleans East, the Lower Ninth Ward, Gentilly and Lakeview.

Among the key policy choices confronted by the city, three stand out as
especially significant (more generally see Popkin et al., 2006; Nossiter,
2006). These are the questions of how to allocate housing assistance funds
to residents, how to restructure public housing, and where to concentrate
support for neighborhood rebuilding. Initial steps favor people and neighborhoods that have more market resources and do little to support the most
disadvantaged.

Housing Assistance

The principal public source of funding for housing reconstruction is the
federal government, with $10.4 billion in Community Development Block
Grant support authorized in June 2006 for Louisiana (Maggi, 2006). The

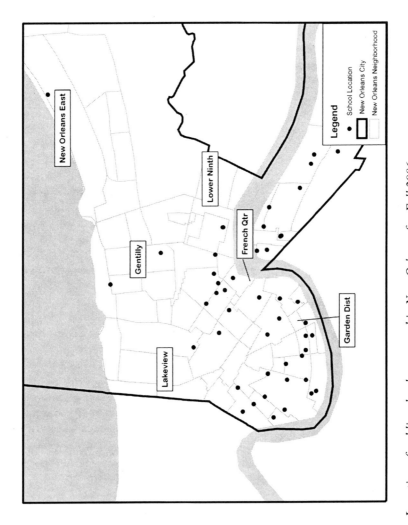

Figure 15.2 Location of public schools reopened in New Orleans for Fall 2006

state's plan for using these funds is called The Road Home Program (Louisiana Office of Community Development, 2006). It targets about $8 billion for assistance to homeowners. Only $1.5 billion is allocated to a program to redevelop rental housing.

According to estimates by the US Department of Housing and Urban Development (2006a) there were a total of almost 100 000 housing units in New Orleans that suffered 'major or severe' damage or were destroyed. Of these 52 percent were owner occupied and 48 percent were occupied by renters. By this measure alone, the policy of allocating most funds to home-owners is not proportionate to the damage. Given that rents have risen by as much as 30–40 percent and the vacancy rate for rental housing is near zero (US Department of Housing and Urban Development, 2006b), it seems clear that a more focused effort to build or rehabilitate rental housing would be needed in order to allow many rental households to return to the city.

Public Housing

A special category of housing is public housing controlled by the Housing Authority of New Orleans. Prior to Katrina there were about 8000 units, though due to poor maintenance (and consistent with a general plan of reducing this segment of the housing stock) only 5100 were occupied at the time of Katrina. As was also true nationally, efforts were under way to restructure the system by demolishing existing units and replacing them with new mixed-income developments. This had already been done with one complex, the St Thomas Project in the Central City and Garden district, which was demolished in 2002 and replaced by a Wal-Mart and new pre-dominantly market-rate condominiums. Originally built for 1500 low-income families, the new development accommodates only 200. Most public housing complexes were sealed after the hurricane (with metal barriers bolted over the doorways) to prevent tenants from returning. Confirming speculation during the preceding year, including comments by public officials that only 'people who are willing to work' should be allowed to return to public housing, specific plans were announced in June 2006 by Housing and Urban Development (HUD) Secretary Alphonso Jackson to demolish 5000 more units (Saulny, 2006). The St Bernard, C.J. Peete, B.W. Cooper and Lafitte housing developments would be entirely removed. They would be replaced by mixed-income developments following the St Thomas model.

At the same time, HUD increased by 35 percent the amount that it would pay as the 'fair market rent' through housing vouchers. If there were vacancies at this rate ($976 for a two-bedroom unit as of October 2006), a system of housing vouchers for low- and moderate-income people could be evaluated as a reasonable alternative to public housing. But in the context of a zero-vacancy rental market at any price, the decision to press ahead with public housing demolition means that few displaced low-income families will have any opportunity to live in New Orleans in the future.

Neighborhood Triage

Although public officials have assiduously avoided saying so, there is a high probability that redevelopment in some neighborhoods will be discouraged by public policy. If this is not done by an overt designation of areas as off-limits for building permits, the same outcome may be achieved passively by choosing not to make the public investments that are required for livable communities. These include repair of infrastructure damage and reopening of facilities such as schools and police and fire stations.

The notion of neighborhood triage was implicit in the proposals made by the Bring New Orleans Back Commission (2005) in the public report of its Urban Planning Committee in January 2006. Some zones of the city were designated as 'immediate opportunity areas' where the city should identify vacant and underutilized property for new construction, expedite permits for repairs and construction of new housing, provide or support community and cultural facilities and services, assist educational and health institutions to address immediate needs, and begin repair and reconstruction using current rules and regulations. Others were proposed to be 'neighborhood planning areas', where the city should conduct a neighborhood planning process to determine the appropriate future. In these areas, the city was advised not to issue any permits to build or rebuild.

Although this report was not adopted, and indeed was widely criticized, it sets forth the key planning question: given that there are insufficient public resources to support fully the rebuilding of all neighborhoods, by what criteria should choices among neighborhoods be made? Mayor Nagin has repeatedly suggested one approach: to let the decisions be guided by residents who 'vote with their feet' to return. Neighborhoods where a will to recover is demonstrated by individual investments and collective action should be supported; in other areas, individuals should be counseled against trying to rebuild. This is essentially a proposal to let the market decide the future of the city.

For many of the same reasons that rebuilding will be facilitated on the Mississippi coast, the white residents of the City of New Orleans are more likely than black residents to be able to return to their neighborhoods, even if the neighborhood is reopened. Whites are more likely to be homeowners (55 percent compared to 42 percent among African American households), but more importantly, they are much more likely to have the personal resources to reinvest in their homes or to find a new residence in a difficult housing market. In the pre-Katrina black population, 35 percent were below the poverty line and the median household income was only $25 000. Among whites, only 11 percent were poor and the median income was more than twice as high – $61 000. Therefore even among homeowners, blacks are less likely to have the means to rebuild than are whites.

In conjunction with public support for homeowners over tenants and the plan to demolish most public housing, market-based policies point toward a future in which New Orleans – though it may be much smaller than before – will also have a smaller share of black residents, tenants, and poor and working-class families. To the extent that the city's labor force continues to require a certain share of persons with low skills and low wages, which is typical of a tourist economy, this means that these workers will mostly live elsewhere.

THE FUTURE IN THE BALANCE: SHIFTING ELECTORAL POLITICS

Although there appears to be an emerging direction in local recovery policy, final decisions remain in flux. The provisions of Louisiana's Road to Recovery and the demolition of public housing are being contested in court, and the election of Democratic majorities in the US House and Senate have so far had little influence on how federal resources will be used. There is also a formal neighborhood planning process put in place in Fall 2006, supported by a large grant from the Rockefeller Foundation, through which neighborhood organizations have been encouraged to make their voices heard. Another factor that may make a difference is the shifting constituencies in local politics.

There has been a potential for political coalitions that cut across the racial and class divisions that have helped structure city politics over the decades. Residents of such very different neighborhoods as Lakeview and the Lower Ninth Ward have a shared interest in short-term assistance programs such as subsidies for temporary housing outside the city. Yet variations across neighborhoods – and across race and class – are likely to support the emergence of a sense of conflicting interests. In December 2005 conflict took the form of opposition to proposals to locate FEMA trailers in public spaces within neighborhoods that sustained less damage. In this case the interests of advantaged neighborhoods (advantaged by protection from flooding and by having residents in place to express their views) were in conflict with the interests of absent residents who have no place to return. Not surprisingly the City Council gave its members veto power over new trailer parks in the areas that they represent.

The mayoral election in spring 2006 offers evidence that conflicting interests are likely to overcome consensus (Logan, 2006b), and it is not obvious whose interests will predominate. On the one hand, there was a substantial change in the composition of the electorate, reflecting the disproportionate displacement of lower income black voters. On the other, there was a surprising shift in the source of Mayor Nagin's electoral support that could make him more responsive to that same constituency.

The first indicator of change is turnout. In the previous municipal election (2002), when the current Mayor, Ray Nagin, defeated Richard Pennington, there was a modest turnout of 130 000 voters (out of a total pool of registered voters that has remained close to 300 000 for the last several years). In the Presidential election of 2004, when few local positions were at stake but there was considerable interest in the contest between President George Bush and challenger John Kerry, turnout was over 197 000. Compared to either standard, participation in 2006 was depressed, with a total of under 115 000 votes cast.

Of course a lower turnout was inevitable. More interesting is how turnout varied across neighborhoods. From the perspective of future urban policy, neighborhoods with the highest electoral participation have likely strengthened their hands in the battles over public investment and development planning that are sure to be a major feature of local politics in the next several years. Figure 15.3 displays turnout levels in 2006 as a proportion of the 2002 level. This map can be compared to the map of racial composition and flood damage presented above.

The neighborhoods with the largest declines in turnout are in the Lower Ninth Ward, New Orleans East, and parts of Mid-City and Bywater. These are all predominantly black neighborhoods, but they have widely varying class composition. In Mid-City and Bywater it is especially the public housing projects whose former residents have been barred from returning to the city up to now. The Lower Ninth Ward is a mixed-income area with many working-class homeowners. These are both areas where the loss of public infrastructure and government restrictions on entry have seriously delayed recovery efforts. New Orleans East, in contrast, has been an important base for the black middle class. All these areas suffered close to 100 percent flooding, and displacement is the most obvious explanation for low turnout.

Among white neighborhoods there is generally a positive correlation between voter turnout and extent of flood damage. For example, the Uptown and Carrollton, and Central City and Garden, districts include some neighborhoods with very little flood damage and others that were hard hit. Neighborhoods with no flooding like Uptown and Garden District had considerably higher participation than in 2002, while those

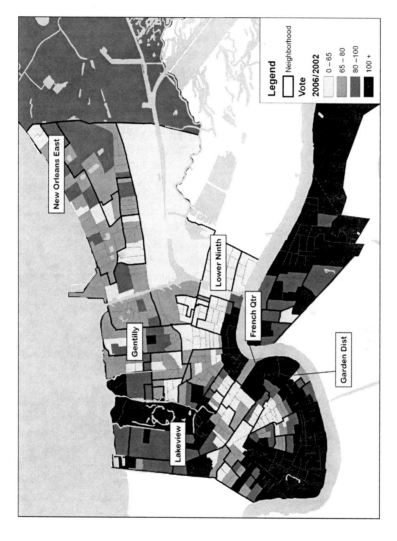

Figure 15.3   Turnout in the mayoral primary election in New Orleans in 2006 in relation to turnout in 2002

with more damage like Broadmoor and Milan suffered a loss. But there are two other significant patterns to point out.

Firstly, several planning districts show little impact of Katrina. The French Quarter and Central Business District actually had higher turnout than in 2002, as did New Aurora and Algiers on the West Bank. These are among the areas of the city with the least flood damage. The surprise here is how much participation declined in comparison to the 2004 Presidential election, with a fall of 25–30 percent that seems unlikely to be due to population loss. In what may have been the most important election in the history of the city, why was turnout in these areas no more than the usual local standard? There may be evidence here of forces beyond displacement, evidence of surprising apathy, alienation and disaffection from the political process by the residents of these relatively advantaged communities.

Secondly, on the other hand, despite its devastation Lakeview shows an exceptional turnout. The number of Lakeview voters was nearly (94 percent) as high in 2006 as in 2002. Even more, there are only modest variations within the district between Lake Shore and Lake Vista, which was only partly flooded, and areas like Navarre that were heavily damaged. Lakeview's participation may have been influenced by a special tax measure on the ballot that would increase property taxes in this district for the purpose of improved policing. A greater factor was probably extensive voter mobilization by local civic groups. Lakeview is known to have a strong civic association that has built upon the many smaller neighborhood associations that used to operate in the area, and in this election it translated its affluence and high levels of homeownership into political clout.

Another planning district with a relatively high turnout despite considerable damage is Gentilly, especially the racially mixed neighborhoods of Fillmore (94 percent as high as 2002) and Gentilly Terrace (88 percent as high).

Although Hurricane Katrina reshaped the political map of the city by suppressing the vote in the poorest and blackest neighborhoods, the dynamics of the mayoral campaign also represent a more remarkable shift in the composition of support for the winning candidate, Mayor Ray Nagin. Having been elected in 2002 on the basis of his strong showing in white and more affluent neighborhoods, the Mayor was re-elected with his main edge among neighborhoods with predominantly black and low to middle income residents.

Figures 15.4 and 15.5 illustrate the shift with maps showing the extent of support for Nagin in his first race in 2002 and in the 2006 primary by precinct. Comparing these maps to the map of racial composition presented above, in 2002 it is clear that Nagin ran strongest in the neighborhoods with smaller black populations. Reports from the period suggested

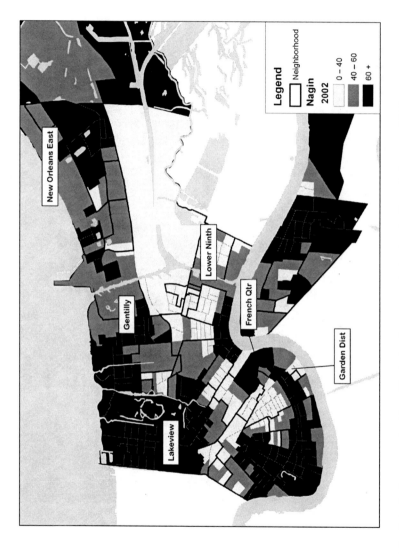

Figure 15.4   Level of support for Ray Nagin in the mayoral election in New Orleans in 2002

293

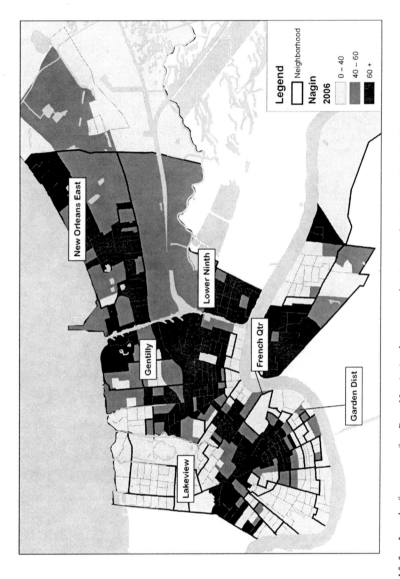

Figure 15.5   Level of support for Ray Nagin in the mayoral primary election in New Orleans in 2006

that in fact his election depended on support from white neighborhoods (and financial backing from people described as the 'Uptown white establishment').

In a surprising remaking of the electoral map, Nagin's support in 2006 shifted almost 180 degrees – neighborhoods that had supported him now supported his opponents, and areas where he had found the least votes now constituted his core constituency. The planning districts with the highest levels of support for Mayor Nagin are the Lower Ninth Ward (83 percent in 2006 compared to only 40 percent in 2002) and New Orleans East (71 percent, up from 55 percent). The individual neighborhoods with the highest shares of Nagin votes are Project neighborhoods (that is, social housing): above 90 percent in Calliope Project, Desire Project and Fischer Project, all areas where he previously received at most a third of votes.

In contrast, Nagin lost heavily in predominantly white areas. In the Garden District, for example, which is only 2 percent black, he gained 85 percent of the votes in 2002 but only 22 percent in 2006. Lakeview, also 2 percent black, voted overwhelmingly for Nagin in 2002 (87 percent) but against him in 2006 (22 percent).

The new political geography of the city will certainly be a factor as policy decisions are made. Where will schools reopen, where will policing and other public services be brought back on line soonest, where will rebuilding be encouraged by city officials and what neighborhoods will have a lower priority? Areas like the Lower Ninth Ward, New Orleans East, and the less affluent and predominantly black sections of Bywater and Mid-City have experienced sharp declines in their participation in the political process. In contrast, Lakeview nearly matched its 2002 vote total despite dislocation of most of its residents, and areas like Algiers, Uptown and Carrollton, the French Quarter and Garden District see their political influence on the rise in this respect.

But there is also a countervailing force, an unexpected consolidation of voting patterns along racial lines in which a politically conservative black mayor turned successfully to a black and low-income constituency that previously had denied him their support. This outcome potentially diminishes the political losses that this part of the electorate, and the neighborhoods where they are concentrated, seemed sure to suffer. Much now depends on how well groups play their cards and what role the backroom players (the investors and real estate entrepreneurs who eventually once again supported Mayor Nagin's campaign war chest) have in the process. Support from areas like the Projects, Lower Ninth Ward and New Orleans East was critical, but their voices will have to be heard from a distance. White Republican neighborhoods backed the losing candidate, but the 20 percent or more of their votes that went to Nagin were indispensable to his victory.

This is a situation where a public official will face conflicting pressures, but may also find considerable room to maneuver and provide leadership for a city that has put key decisions on hold for too long.

REFERENCES

Bring New Orleans Back Commission, Urban Planning Committee (2005), 'Action plan for New Orleans: the new American city', accessed 7 January 2006 at http://www.bringneworleansback.org/Portals/BringNewOrleansBack/Resources /Urban%20Planning%20Final%20Report.pdf.

Colten, Craig (2005), *An Unnatural Metropolis: Wresting New Orleans from Nature*, Baton Rouge, LA: Louisiana State University Press.

Gabe, Thomas, Gene Falk, Maggie McCarty and Virginia W. Mason (2005), 'Hurricane Katrina: social-demographic characteristics of impacted areas', Washington, DC: Congressional Research Service, Library of Congress, accessed 6 January 2007 at http://www.gnocdc.org/reports/crsrept.pdf.

Logan, John R. (2006a), 'The impact of Katrina: race and class in storm-damaged neighborhoods', Report by the American Communities Project, Brown University, Providence, PI, 25 January, www.s4.brown.edu/Katrina/report.pdf.

Logan, John R. (2006b), 'Population displacement and post-Katrina politics: the New Orleans mayoral race, 2006', report by the American Communities Project, Brown University, Providence, PI, 1 June, www.s4.brown.edu/Katrina/report2. pdf.

Louisiana Department of Health and Hospitals (2006), 'Louisiana health and population survey report', 29 November, Orleans Parish, accessed 6 January 2007 at http://popest.org/popestla2006/files/PopEst_Orleans_SurveyReport.pdf.

Louisiana Office of Community Development (2006), 'Action plan for the use of disaster recovery funds', accesssed 6 January 2007 at http://www.doa.la.gov/ cdbg/dr/plans/ActionPlan-Approved_06_04_11.pdf.

Maggi, Laura (2006), 'Housing aid plan challenged: renters neglected, nonprofit groups say', *Times-Picayune*, 21 June, p. 2.

Nossiter, Adam (2006), 'In New Orleans, money is ready but a plan isn't', *New York Times*, 18 June, p. 1.

Popkin, Susan, Margery A. Turner and Martha Burt (2006), 'Rebuilding affordable housing in New Orleans: the challenge of creating inclusive communities', Washington, DC: Urban Institute, accessed 7 January 2007 at http://www. urban.org/UploadedPDF/900914_affordable_housing.pdf.

Russell, Gordon (2006a), 'Address changes offer insight into city', *Times-Picayune*, 5 February, p. 1.

Russell, Gordon (2006b), 'Six months later, recovery gaining focus: city may be near turning point', *Times-Picayune*, 26 February, p. 1.

Saulny, Susan (2006), '5000 public housing units in New Orleans are to be razed', *New York Times*, 15 June.

US Department of Housing and Urban Development, Office of Policy Development and Research (2006a), 'Current housing unit damage estimates: Hurricanes Katrina, Rita, and Wilma', 12 February revised 7 April, accessed 7 January 2007 at http://www.gnocdc.org/reports/Katrina_Rita_Wilma_Damage _2_12_06___revised.pdf.

US Department of Housing and Urban Development, Office of Policy Development and Research (2006b), 'Economic and housing market conditions of the New Orleans–Metairie–Kenner, Louisiana Metropolitan statistical area as of September 1, 2005, with updates to February 1, 2006', accessed 7 January 2006 at http://www.huduser.org/publications/pdf/CMAR_NewOrleansLA.pdf.

Index

ability to pay 220, 222
access management plans 68
adaptive resilience 200, 201
adjustment
    to disaster shocks 236–9, 248
    to natural disasters 5–6, 208–27
Advanced Circulation Model for
    Coastal Hydrodynamics
    (ADCIRC) 57–8
age
    evacuation by 105–10
    perceived risk by 101–4
    pre-hurricane preparation by 104–5
agency costs 127
agent-based modeling 94
agriculture 236, 237, 238, 239
air transport 81, 131
    New Orleans 108, 112, 134–9,
    142–5
AIR Worldwide 17
all-hazards approach 2–3, 23–30,
    71–82
Allstate 11
American Airlines 108
analysis submodel 38, 40
Analytica 36, 37, 42
asset prices 227
assignment models 3, 93
Astrodome (in Houston) 108
Atchafalaya River 38, 40, 44, 47, 48
attacker
    best response 74, 75–6, 79
    effort 2–3, 71–9, 80, 81–2
    games 74–5, 77–8
    repeated attacks 81
aviation security 81

back-up supplies (water) 199, 201–2
backflow 66, 67
backswamp areas 255, 256
Bangladesh 86, 235, 240, 242–5
'boil water notice' 188, 196–8, 205

Bondi, Sir Hermann 4, 120, 125–9,
    131–2
border security 72
bounded rationality 203
Box-Cox transformation 148
breach 37–40, 43, 50, 60–66, 70
bridges 68, 70
Bring New Orleans Back Commission
    288
Brookings Institution 268
'budget aggregate' approach 194–5
building inspection (Dade County)
    214, 227
building standards/code 36, 269–70,
    288
Bulletin 17 40–41
business
    interruption 5, 10, 123, 127, 187, 188
    multi-sector model 190–92

Caisse Centrale de Reassurance (CCR)
    24–5
California Earthquake Authority
    14–17
Calliope Project 295
'candle with matches' 240–41
capital 127, 132, 237, 238
    see also human capital; physical
    capital
cascading effect 94, 116
Caribbean island states 237–8
casino gambling 4, 284
catastrophes
    household production function
    203–5
    insured 9–13
    see also disasters; natural disasters
causality, insurance and 25–6
Chenery-Moses type model 148
China 235
Churchill Archives Centre 126
circular flow of economy 190–91

'Citizens Corps' program 116
Citizens Property 22–3
class 7, 279–80, 282, 284, 289, 290
climate change 11, 236, 239, 249
Cobb-Douglas utility function 192
combined ratio (insurance) 13–14
Commodity Flow Survey 148–9
communication systems 68, 115, 116
Community Development Block
    Grants 269, 285
comprehensive disaster insurance 8–30
computable general equilibrium (CGE)
    models 188, 190, 192–4, 198–9,
    204
computational simulation modeling 3,
    94
Congressional Research Service 253,
    280
conservation (of water) 201–2
Consorcio de Compensación de
    Seguros 24
constant elasticity of substitution
    (CES) 192, 196, 197–8, 201
Consumer Prices Indices (CPI) 44
consumers 28, 200
    behaviour 188–9, 192–9, 203–5
consumption 188, 190–91, 192, 248
contraflow loading plans 68
Convention Center 67, 100, 108–9, 279
convex disutility of effort 79, 80–81
coping mechanisms 5, 187–205
corruption 132
cost-benefit analysis 2, 95, 96, 131, 235
costs 46, 199
    agency 127
    capital 127, 132
    of floods 47, 49, 123
    housing rebuilding 268–75
    maintenance 122, 127, 128, 132
    mitigation 47, 48
    redevelopment 7, 272–4
    Thames Barrier 121–2, 127–30, 131,
        132
    transport 152, 189
    unit cost function (of CES) 197–8
counterterrorism 81
credit 204, 238
Creole Cottage/Creole Townhouse
    255–6
crisis communications 116

culture 84–5, 86, 208
cumulative distribution function 42–3,
    45
cumulative severity distribution 44,
    46–7
Current Population Survey 283
curtainwall failures 56
Customs District of Louisiana 147,
    152, 153, 160–74
cut-offs 44–5, 47, 49–50

Dade County (Florida) 6, 209–27
damage 9, 56
    adjustment following 5–6, 208–27
    assessment 7, 253, 258, 261–7
    disutility of 77, 80
    flood 261, 263–5, 266, 280–81, 284
    housing 5, 7, 204–5, 215, 238, 256,
        258–67
    market response 210–13, 223–6
    to neighborhoods 258, 261, 262
    to New Orleans 5, 204–5, 256,
        258–67
    policy choices 279, 284–9
    probability 71, 72, 73
    utility of 72, 73, 75–6
    wind 213–14, 253–4, 279
    see also insurance
data needs 198–9
deaths 9, 12, 53, 60, 86, 89, 107, 115,
    123, 208
    economic value of lost lives 46–7,
        48–9
decision analysis 2, 34–51
decisions
    evacuation 3–4, 93–117
    risk assessment 93–4, 95–7
decontamination (of water) 202
defender
    best response 76–7
    defensive investment 2–3, 71–80, 81
    first-mover advantage 72, 74
    games 74–5, 77–8
defensive strategies 71, 72, 74, 81
Delta Project 241–2
demand-side NIEMO 147, 149–50,
    159
demographics 6, 37–8, 209, 211–23
Department for International
    Development (DFID) 235

design errors 37, 38, 39, 42, 43
designated floodplains 36, 50, 102
Desire Project 295
desired state 199, 201
deterrence 81
developing countries 6–7, 230–49
Dhaka 240, 243, 245
direct impacts 152–8, 160–61, 164–5,
   179–86
'dirty side' of storm 253
disaster risk
   longer-term development 235–9
   recognition 240, 243–7
   reduction 6, 230–32, 234–5, 238–47,
      248
disaster shocks 232–9, 248–9
disasters
   cascading effect 94, 116
   exposure 231, 232, 234, 236, 237,
      238–9
   institutional 88–90
   macroeconomic consequences 232–9
   regional impacts 5, 187–205
   response/recovery 54–5, 201–3
   see also catastrophes; insurance;
      natural disasters; terrorism
disease 115
displacement 7, 279–84, 290
disutility 72–3, 75–7, 79–81
diversification 234
Docklands Light Rail (DLR) 123
documentation (of results/lessons)
   113–17
dollar expenditure (insurance) 19
Dominica 245, 246, 247
double-log distance-decay function 148
Double Gallery House 256
doubly constrained Fratar model 148
drainage 69, 243–4, 284
Draper, Charles 125
drought 24, 233, 234
Duke University 94

earned premiums (insurance) 13, 14
earthquakes 234
   insurance 14–15, 16–17, 24, 25, 27
   Northridge 5, 10, 12, 14, 199, 204
East Pakistan Water and Power
   Development Authority
   (EPWAPDA) 244

Eastern Caribbean 243, 245–7, 249
economic adjustment 236–9, 248
Economic Commission for Latin
   America and
   the Caribbean (ECLAC) 233
economic disruption 9, 200–201
economic impact
   of Hurricane Katrina 5, 147–75,
      178–86
   regional 5, 187–205
economic inflation adjustment 44
economic losses 5, 187–205
economic performance 233–5
economic recovery 4–5, 134–45
economic value of lives lost 46–7, 48–9
economies of scale 200
economy, circular flow of 190–91
education 209, 217, 218, 226, 285, 286
effort 2–3, 71–82
El Niño 233
Elbe floods 7
electrical power 2, 66, 67, 116, 285
electoral politics 289–96
elevation (New Orleans) 254–6
elite panic 90
Emergency Management Agencies 116,
   117
emergency plans 69–70
emergency responders 3, 111, 115–16
Emergency Response Councils 116
emergency shelters 106, 107, 204, 276,
   283
employment 135, 136
   see also unemployment
endogenous attacker effort 2, 71, 81
endogenous growth 237–9, 248
engineering design 60–67, 69
Entergy Corporation 285
environment 90, 203, 222
Environment Agency 122, 125, 129,
   130–31
equilibrium
   CGE models 188, 190, 192–4, 198–9,
      204
   defensive strategies 2, 81
   general 190, 200
   Nash 74
   pure-strategy 79
   sequential game 2, 78
   simultaneous game 2, 77–8

equity (role of insurance) 20–23, 24, 29
erosion failures 61–2, 63, 64–5, 70
Essence Festival 143
Europe (insurance sector) 24–5
European Space Research
    Organization 125
Europort 128
evacuation 89
    by age/race 105–10
    decisions 3–4, 93–117, 203–4, 209
    displacement pattern 7, 279–84,
        290
    lessons learned 66, 67–8
    mandatory 4, 93, 95, 106, 108, 109,
        110, 145
    modes *see* transport
    plans 85, 115
    routes 93, 106, 107, 115, 116
    time 46, 48, 49–50, 67
exceedence probability 41, 42
executive failure 90, 91
exercises (emergency management) 117
expecting the unexpected 117
exogenous attribute 211
exogenous change 211, 212, 226
exogenous growth 237
exogenous shocks 217, 227, 235
experience, previous 95–7, 101, 102,
    131
exports
    actual/estimated/forecast 153–5
    direct impacts 152–8, 160–61, 164–5,
        179–82
    domestic/foreign ratio 152–3, 178
    losses 147, 150, 159–61, 164–5,
        168–70, 174, 179–82
exposure 231, 232, 234, 236–7, 238–9
externality/filtering models 211–12
'extraordinary risks' (in Spain) 24
extreme value distribution 41

factor of safety 63–4, 65
failure
    executive 90, 91
    institutional 230
    -investigation cycle 70
    market 237, 238
    official 3, 84–92
    response 90
family evacuation plans 109, 111, 115

Federal Emergency Management
    Agency (FEMA) 7, 36, 108, 115
    flood maps/zones 57, 209–10, 212,
        216, 221, 223–6
    housing and 258, 262–4, 269–70,
        272, 274, 283, 289
    Hurricane Andrew 209–10, 212, 216,
        220–21, 223–6
    worst-case scenarios 87, 90–91
fire damage 25–6
'first responders' 111
fiscal management 235
Fischer Project 295
flood
    consequences (evaluation) 44–6, 47,
        48
    costs 47, 49, 123
    damage 261, 263–6, 280–81, 284
    disaster response (New Orleans)
        54–5
    frequency risk analysis 39–44, 45
    housing and 102, 103–4, 253–4
    insurance 11, 14, 15, 17, 20–21,
        24–8, 36, 89–90
    maps/zones 57, 209–10, 212, 216,
        221, 223–6
    pump stations 2, 54, 65–7, 69, 70,
        284
Flood Commission (Pakistan) 244
flood control
    Bangladesh 240, 242–5
    decision analysis 2, 34–51
    protection levels 55–7
    redundancy within schemes 241–3
    Thames Barrier 1, 2, 4, 120–32
floodplain 69
    damage assessment 261, 263–6
    designated 36, 50, 102
floodwalls 3
    design 35–6, 61
    lessons learned 54, 59–61, 62–5,
        68
    rebuilding 34, 35–45, 49–50
Florida 16, 22–3
    Dade County 6, 209–27
Florida Hurricane Catastrophe Fund
    16, 23
food-for-work programs 244
France (insurance sector) 24, 25
frequency risk analysis 39–45, 239

funding 70
  defensive investment 2–3, 71–80, 81
  see also subsidies

Galveston Hurricane 86
game theory 2–3, 72–8, 79–81, 82
gap formation 63–4, 65, 70
Garden District 99, 100, 255, 282
gas industry 5, 147
general equilibrium 190, 200
geographical weighted regression 149
geophysical hazards 234, 236
Georgia State University 23
Ghoshian price model 151–2
global positioning system (GPS) 97
global warming 130
Great Flood (1953) 123–5
Great Mississippi Flood (1927) 39, 54
Greater New Orleans 254–5, 258,
  259–60
greenhouse gas emissions 4
Greenwich 129
gross domestic product (GDP) growth
  232–3
Gulf Coast 25, 26, 29, 54, 84
Gulf Intracoastal Waterway 58, 59, 61

hazard-prone areas 9, 18, 20–21, 24,
  29
Heathrow Airport 131
hedonic models 213
highway bridge decks 68
Hollandse Ijssel 241–2
Holt-Winters time-series approach 153,
  155, 157
Home Builders Association 269
home ownership
  adjustments 209–11, 215, 217–18,
    220–24
  damage and 215, 261–7, 272, 284–5
  impact of Hurricane Katrina 254,
    258, 261–7
  insurance and 14, 18, 21–8, 89–90
homeland security 8, 29, 82, 91, 108,
  116, 126, 159, 258
Homestead Air Force Base 212,
  220–21
hospitals 135, 136
hotels 4, 135, 136, 139–41, 142–5
House Document No. 465 15

household production function (HPF)
  theory 188, 189, 190–99, 201,
  203–5
households
  impacts of disasters 5, 187–205
  income 209–13, 216–26, 227
  multi-sector model 190–92
  production technologies 193–5, 203
  resilience 199–203
  utility function 194, 196
  utility lifelines 5, 187–205
housing
  adjustments 6, 209–10, 212–20,
    223–7
  assistance 285, 287
  below sea level 102, 103, 104
  damage 5, 7, 204–5, 238, 256, 258–67
  in flood zone 102, 103–4, 253–4
  geography of New Orleans 254–60
  impact of Hurricane Katrina 204,
    253–76
  permits 136
  Projects 280, 282, 287, 290, 295
  prospects for recovery 258, 261–7
  public 280, 282, 285, 287–90
  rebuilding 36, 265–74, 288
  rescue of residents/visitors 110–12
  tenure *see* housing tenure
  types 255–6
  units (by age) 256, 257
  values 209–10, 213, 216, 219, 220,
    223–5
  vouchers 275, 287
Housing Authority of New Orleans
  287
housing tenure
  damage by 261–7, 272, 284–5
  see also home ownership; public
    housing; rented housing
human capital 147, 235, 236, 237–8
Hurricane Andrew 10, 11, 12, 15–18
  adjustments 5–6, 208–27
Hurricane Betsy 15, 35, 54–5
Hurricane Camille 35, 84
Hurricane Charley 12
Hurricane Frances 12
Hurricane Georges 12
Hurricane Hugo 10, 12, 246
Hurricane Ivan 12, 102
Hurricane Jeanne 12

Hurricane Katrina 131–2, 208
  aftermath 2–3, 71–82
  conditions prior 53–7
  as defining event 6–7, 230–49
  economic impact 5, 147–75, 178–86
  effects on New Orleans 57–60
  impact on housing 7, 204, 205,
      253–76
  institutional disaster 89–90
  insurance 12, 17, 20, 25–6
  lessons learned 1, 2, 53–70
  rebuilding flood controls 3, 34–51
  resilience 5, 187–205
  risk (challenges) 3, 93–117
  social impacts 7, 279–96
  timeline events 107–8
  tourism 4–5, 134–45
  worst-case thinking 3, 84–92
'Hurricane Pam' exercise 117
Hurricane Rita 12, 107, 108, 113,
    168–74, 232, 283
Hurricane Wilma 12
hurricanes 8–9, 11
  consequences (evaluation) 44–6, 48
  disaster response 54–5
  insurance 15–16, 17–18, 20, 22, 23,
      25, 27–8
  protection system (HPS) 55, 60–62,
      65–6, 69–70, 243
  rebuilding flood controls 35–6,
      38–41, 46, 48
  Saffir-Simpson scale 34, 40, 41, 55–6

I-walls 62, 63, 64, 65, 70
imagination (worst-case scenarios) 3,
    86–7
IMF 233, 235
'immediate opportunity areas' 288
IMPLAN input-output models 148–9
imports
  actual/estimated/forecast 152–3,
      156–7
  direct impacts 152–8, 162–3, 166–7,
      183–6
  losses 147, 150, 159, 162–3, 166–7,
      171–4, 183–6
Incident of National Significance 108
income
  adjustment and 6, 208–13, 215–16,
      218–23, 225, 226, 227

displacement and 280, 282, 283–4
household 209–13, 216–26, 227
housing and 287–9, 290
low (protection) 9, 20–21, 22
mean 211
median 6, 210, 220, 222, 284, 289
voting behaviour and 292, 295
  see also poverty
Independent Levee Investigation Team
    (ILIT) 60–61, 63–4, 65, 243, 248
Indian Ocean tsunami 208, 231, 249
Industrial Canal 108, 258, 280
inflation 44
influence diagrams 2, 35, 36–7
information 57, 231, 246, 248–9
  on risk 208, 210–13, 220–26, 227
infrastructure 8, 35, 44, 69, 84–5, 116,
    189, 238, 288, 290
inherent resilience 200, 201
Inner Harbor Navigational Canal
    (IHNC) 58, 59–60, 61, 63, 65
input-output models
  multi-state 5, 147–75, 178–86
  resilience and 188, 190, 192–4, 197
institutional disasters 88–90
institutional failure 230
insurance 232, 238, 246, 274, 285
  adjustment and 6, 220, 222, 227
  affordability 20–23, 24, 29
  combined ratio 13–14
  companies 10–11, 27–8, 247
  comprehensive (role) 1–2, 8–30
  coverage 26, 28–9
  of extreme events 9–14
  flood 11, 14, 15, 17, 20–21, 24–8, 36,
      89–90
  future programs 17–23
  mandatory 16, 20–21, 28–9
  risk-based all-hazards 23–9, 30
  risk-based premiums 18–20, 23, 27,
      28
  US system 14–17
  wind 15–16, 24–5
Insurance Information Institute 10, 12,
    14, 17, 23
Insurance Service Office (ISO) 17
insured losses 9–14, 18, 27
integrated coastal zone management
    50–51
intelligence (counterterrorism) 81

intentional threats 73, 80, 81, 82
Interagency Performace Evaluation
    Task Force 34, 51, 59–65, 231,
    239–40, 248
intermediate goods 190, 191
International Building Code 269
International Panel on Climate Change
    11
Interregional Input-Output model
    148
interviews (and data) 97–112, 114
investment
    defensive 2–3, 71–80, 81
    economic adjustment 237, 238, 248
    housing 265, 268, 270–71, 274, 288,
        290
    public 274, 288, 290
Iraq 80, 114
irrigation 243–4
Israel 80

Japan 147, 208
Jazz and Heritage Festival 5, 143

knowledge, level of 114–15
Kobe earthquake (Japan) 147
Krewes 141–3, 145
Krug Mission 244

labour
    human capital 147, 235, 236, 237–8
    productivity 189, 192
Lackland Air Force Base 108
Lake Pontchartrain 110, 256
    flood frequency modeling 40–43,
        47–9
    submodel 37–8, 39
land
    subsidence 38–9, 42, 43, 50, 55, 69
    use 35, 37, 38, 220, 231, 239, 272
Land and Water Sector Study 244
leisure 188, 197, 198
    see also tourism
Leontief models 150–51
levee systems
    failures 60–61, 63–5, 117, 254
    ILIT 60–61, 63–4, 65, 243, 248
    lessons learned 1, 2, 53–70
    rebuilding 1, 2, 34–51, 268–70, 276
    transitions 61, 62–3

life
    economic value of 46–7, 48–9
    lost lives 46–7, 48–9, 234
    -safety systems 61, 69–70, 204
linear expenditure system 192, 196–7, 199
log-Pearson Type III analysis 41, 43
logistic distribution 41–2
lognormal distribution 41
London Avenue Canal 63, 64–5, 108
London transport system 80
losses
    cumulative 44, 45, 46, 47
    economic (resilience) 5, 187–205
    economic (state-by-state) 5, 147–75,
        178–86
    export 147, 150, 159–61, 164–5,
        168–70, 174, 179–82
    household 188, 189
    import 147, 150, 159, 162–3, 166–7,
        171–4, 183–6
    insured 9–14, 18, 27
    lost lives 46–7, 48–9, 234
Louis Armstrong International
    Airport 108, 112, 136–7
Louisiana
    housing damage 26, 253, 269, 272–4
    see also New Orleans
Louisiana Citizens Property Insurance
    Corporation 274
Louisiana Customs District 147, 152,
    153, 160–74
Louisiana Department of Health and
    Hospitals 279
Lousiana Office of Community
    Development 287
Louisiana Recovery Authority 7
Louisiana Road to Recovery 289
Louisiana State University 57
Loyola University 282

macroeconomic consequences of
    natural disasters 232–9
Madrid train explosion (2004) 80
maintenance costs 122, 127, 128, 132
Maldives 234
man-made hazards (regional economic
    impacts) 5, 187–205
mandatory evaluation 4, 93, 95, 106,
    108, 109, 110, 145
mandatory insurance 16, 20–21, 28–9

Mardi Gras 4–5, 134–6, 141–5
marginality 6, 230
market(s)
    failures 237, 238
    goods 188, 192, 193–5, 197, 203
    residual 20, 22–3
    responses (to Hurricane Andrew)
        210–13, 223–7
Marshlands (restoring) 68
materials, improper 37, 38, 39, 42, 43
mayoral elections 292–6
media 109, 110, 116
megacities 234, 243
Meteorological Office 122
Metropolitan Service Area (MSA) 135
Miami Herald 209, 212–14, 216,
    220–21, 223–4
Mid-City districts 270–71, 273
Millennium Development Goals 233
minimum surface pressure 40–41, 56
Ministry of Agriculture, Fisheries and
    Food 121
missing persons 12, 115
Mississippi (wind damage) 253–4
Mississippi River 256, 284–5
    flood frequency modeling 36–40,
        42–3, 45–6, 47–8, 49–50
    flood submodel 37, 38
    Gulf Outlet (MRGO) 58–9, 61, 258
mitigation costs 47, 48
mitigation measures 116, 147, 200, 268
    incentives 19–20, 29
mixed-income developments 287
mobility characteristics 95
Monte Carlo sampling 38
Montserrat 234, 245, 246–7
moral hazard 239
multi-sector model 190–92
multi-state input-output model 5,
    147–75, 178–86
multi-target games 73, 79–81, 82
multiplier effects 147, 149, 159, 190
multiregional input-output (MRIO)
    model 5, 147–75, 178–86
Munich Re 13

Nagin, Mayor Ray 292–6
Nash equilibrium 74
National Commission on Terrorist
    Attacks Upon the United States 71

National Flood Insurance Program
    (NFIP) 14, 15, 17, 20, 21, 24, 26,
    36
National Hurricane Center 105
National Interstate Economic Model
    (NIEMO) 5, 147
    background 148–9
    elaborations 150–52
    estimation of direct impacts 152–8
    previous applications 148–50
    results 159–75, 178–86
    three-ports applications 149–50
National Oceanic and Atmospheric
    Administration (NOAA) 44, 209,
        210, 213, 214, 216, 220–21, 222,
        224–5
National Science Foundation 94
*National Strategy for Homeland
    Security* 8
National Weather Service 44, 55, 57
natural disasters
    adjustment to 5–6, 208–27
    insurance *see* insurance
    insured losses 9–14
    Katrina as defining event 230–49
    macroeconomic consequences 232–9
    optimal protection 2–3, 71–82
    regional economic impacts 5,
        187–205
'neighborhood planning areas' 288
neighborhoods
    damage to 258, 261, 262, 280–81
    policy choices 7, 279–81, 288–9
    rebuilding decisions 265–7, 269–72
    triage 288–9
net annual flood costs 47, 49
Netherlands 241–2, 249
New Orleans
    damage 5, 204–5, 256, 258–67
    displacement from 7, 279–84, 290
    economic impact of Hurricane
        Katrina 147–75, 178–86
    effects of Hurricane Katrina 57–60
    future (electoral politics) 289–96
    geography of 254–6, 257
    housing sector 7, 253–76
    lessons learned 2, 53–70
    official failure 3, 84–92
    population data 254, 256, 259
    reconstruction 68–9

recovery (data) 135–6
recovery (prospects) 258, 261–7
risk (decisions) 93–117
social impacts/policy choices 7,
    279–96
tourism 4–5, 134–45
worst-case scenarios 3, 84–92
New Orleans Airport 137
New Orleans flood control system 2
    evaluation of consequences 44–6, 47,
        48
    flood frequency risk analysis 39–44,
        45
    floodwalls/levees system 34, 35–6
    model overview 36–9
    preliminary results 46–8, 49
New Orleans Metropolitan Service
    Area 135
9/11 Commission Report, The 71–2
NOLA Flood Control Risk Analysis
    System 36–7
non-devastated areas (NDAs) 270–73
non-intentional threats 73, 80
non-market costs 199
non-market goods 188–9, 192, 203, 205
non-market responses (to Hurricane
    Andrew) 210–13
Northridge earthquake 5, 10, 12, 14,
    199, 204

Office of Insurance Regulations 23
oil industry 5, 147, 256
Old River Control Structure 47
'on the fly' plan 106
one-target games 72–3, 75–8, 80, 81,
    82
opportunity cost
    of attack 72
    of household labour 190, 191
    of leisure time 188, 197
output 5, 147–75, 178–86
overtopping 37, 38, 50, 58–9, 60, 61–2,
    66, 70
owner-occupiers *see* home ownership

Palm Beach Post 23
panic 90
parametric analysis 34, 35
passive survivability 275–6
'peak annual flood discharges' 41

people
    culture and 84–5
    see also victims
performance
    economic 233–5
    evaluation *see* Interagency
        Performance Evaluation Task
        Force (IPET)
personal evacuation plans 115
pessimistic approach 91–2
physical capital 147, 235, 236, 237–8
piping 64–5
planning districts 258, 264–7, 270,
    280–82, 292, 295
planning failures 90
Poe Inc. 11
Poisson distribution 42, 43
policy choices (after Hurricane
    Katrina) 7, 279–96
political indifference 89
politics (influence) 279, 289–96
population
    changes 44, 113, 114, 135, 136
    density (effect of bombing) 208–9
    growth (Dade County) 214
    New Orleans 254, 256, 259, 280
    reduction 85
Port of London 123, 127, 128
    Authority 122
possibilistic approach 3, 87–8, 91
postcodes 99, 100, 101, 282–3
poverty 89, 265
    developing countries 6, 230, 234–5,
        239
    policy choices 280, 282, 285, 289
'pre-emptive resilience' 115
preparation 3–4
    evacuation (by age/race) 105–10
    interviews (data) 97–101
    lessons documented 113–17
    perceived risk and 101–4
    population changes 113, 114
    pre-hurricane (by age/race) 104–5
    rescue of residents/visitors 110–13
    risk assessment 93–4, 95–7
price model 151, 152
prices 44, 211, 237
    housing 223, 227, 268, 275
    shadow 192, 193, 197, 198, 203
PricewaterhouseCoopers 15, 16

probabilism 87, 248, 249
probability 3, 19, 26, 41, 50, 91, 127,
    131
  cumulative 46, 47
  of damage 2, 71, 72, 73
  density function 42
production 188, 195, 200
production function 151
  household 188, 189, 190–99, 201,
    203–5
productivity shocks 236, 237, 238, 239
Project neighborhoods 280, 282, 287,
    290, 295
property damage *see* damage
protection, future (Thames Barrier)
    129–31
protection, optimal 3, 71
  assumptions 73–4
  games 2, 72–82
public expenditure reviews 235
public goods/bads 220
public housing 280, 282, 285, 287–90
pump stations 2, 54, 65–7, 69, 70, 284
pure-strategy equilibrium 79

quantity model 151, 152
quasi-experimental analysis 223–6, 227

race
  adjustment by (Dade County) 215,
    216, 217–18, 220–21, 222, 226
  displacement and 7, 128, 279–84
  evacuation by 105–10
  home ownership and 288–9
  perceived risk by 101–4
  pre-hurricane preparation by 104–5
  voting behaviour 292, 295
rail transport 128
Raised Centerhall Cottage 256
RAND Corporation 20
rebuilding
  decisions/policies 265–7
  redevelopment costs 272–4
  strategies 267–74
recovery, economic 4–5, 134–45
redevelopment costs 7, 272–4
redundancy, DRR and 231, 239–40,
    241–3
regional economic impacts of hazards 5
  conceptual framework 190–92

household activities 187–9
household production function
    192–9, 203–5
household resilience 199–203
Regional Transit Authority 106
reinsurance market 23, 24, 27
'relief values' 47
rented housing
  adjustment and 208, 210, 212–13,
    216–26
  damage 215, 258, 261–7, 285
  public 280, 282, 285, 287–90
rents 210, 212–13, 216, 219–20, 222–5,
    226, 275
repeat sales analysis 226, 227
rescue 4
  emergency responders 3, 111, 115–16
  of residents/visitors 110–12
residential patterns 89
residents, rescue of 110–12
residual markets 20, 22–3
resilience 5
  conceptual framework 190–92
  DRR 231, 232, 238, 240–41
  household 187–9, 199–203
  household production function
    192–9, 203–5
resource allocation 126, 248
  defensive investment 2, 71–80, 81
  optimal protection 2–3, 71–82
resource scarcity 2, 151, 238
response
  failures of 90
  market 210–13, 223–7
  non-market 210–13
restaurants 4, 136
'return period' 41
reverse flow 66, 67
riots 24
risk 69
  assessment 6, 95–7, 230, 234, 239,
    247, 276
  attitude 79–80
  -based all-hazards 23–9, 30
  -based premiums 18–20, 23, 27, 28
  -cost analysis 126, 132
  disaster (longer-term development)
    235–9
  DRR 6, 230, 231–2, 234, 235,
    239–47, 248

diversification 25, 27
frequency analysis 39–44, 45, 239
information 208, 210–13, 220–26, 227
perceived (by age/race) 101–4
profile 3–4, 93–4, 95–7
recognition 240, 243–7
rescue and 3, 110–12, 115–16
transfer instruments 27, 28
Risk Management Solutions 16
riverine floods 7, 234, 240, 243–5
Road Home Program 273, 274, 287
Road to Recovery 289
roads/road transport
London 4, 128
New Orleans 106, 107
Rockefeller Foundation 289

sabotage 35, 42, 43
safety 61, 63–5, 69–70, 204
Saffir-Simpson scale 34, 40, 41, 55–6
St Thomas Project 287
sales tax collection 136, 145
San Francisco earthquake 25–6
scarcity 151, 238
schools 238, 285, 286
scour 62
Sea, Lake and Overland Surges from
    Hurricanes (SLOSH) model 57
sea level
London 130
tidal surges 120, 122–3, 126, 127–8,
    130, 132
sea transport/trade 125, 127–8, 147,
    149–50, 159
search costs 152
Section 8 housing 282
seismic hazards 234, 245–7, 249
    see also earthquakes
Seismic Research Unit (SRU) 246, 247
self-protection 5, 6, 209, 220, 222, 227
sensitivity, risk and 232, 234, 237
September 11 (2001) 5, 8, 10, 11, 12,
    91, 115, 144
aftermath (protection) 2–3, 71–82
sequential games 2, 72, 74–5, 78, 79
17th Street Canal 63–4, 108
shadow prices 192, 193, 197, 198, 203
shear failure/resistance 63–4, 70
Sheerness-Southend barrier 124, 129,
    130

sheetpile curtains 63
shelters 106, 107, 204, 276, 283
Shotgun House 256
simultaneous games 72, 74–5, 76–8, 80
single-state subsidiaries 28
single-target game 72–3, 75–8, 79,
    80–82
Small Business Administration 258,
    262–4, 269, 272–3
small insurers 27–8
small island economies 234
Smith Travel Research 136
social exclusion 6, 230
social impacts (of Katrina) 7, 279–96
social interaction model 212, 216, 222
socio-economic factors 95
soil erosion 61–2
soil failure 70
Solow-Swan growth model 237
sorting models 211–12, 213, 217, 219,
    227
Southeast Hurricane Disaster Relief
    Act 15
Southwest Airlines 138
Spain (insurance sector) 24
spatial differentiation 208
Special Flood Hazard Areas 17
stability failures (levees) 61, 63–5
staged evacuation plans 68
'stages of life' characteristics 101
standard project hurricane 42, 54, 55
state-by-state economic impacts (of
    Hurricane Katrina) 5, 147
direct impacts 152–8, 179–86
import/export losses 159, 168–74
NIEMO (applications) 148–52
NIEMO results 159–75
State Board of Administration of
    Florida 16
state of emergency 105
State Farm Fire and Casualty 10–11
steady-state capital 237
Stern Review 236, 239
Stolper-Samuelson model 238
Storm Tide Forecasting Service 122
storms 11, 12, 22, 24–5
categories 40–41
'dirty side' 253
surge barriers (Netherlands) 241–2
surges 55–60, 63, 66, 241–2, 253–4

strategic defense 71, 72, 74, 81
sub-Saharan Africa 233, 236
subgame-perfect Nash equilibrium 74
subsidence 38, 39, 42–3, 50, 55, 69
subsidies
    housing rebuilding 261, 268, 269,
        273–4
    housing rents 273, 274, 282, 285,
        287, 289
    insurance 15, 17, 20–21, 27, 29
substitutability (inputs) 202
substitutes 152, 159, 192–3
substitution effects 188, 192
    CES 192, 196, 197–8, 201
'super-cats' 9
Superdome 67, 100, 106, 107, 108–9,
    276, 279
supply-side (of NIEMO) 147, 149–52,
    159
survival, resilience and 204, 205
sustainable development 239
Swiss Re. 10, 11, 12

T-walls 62, 65
targets
    attractiveness 72
    multi-target games 73, 79–81, 82
    value of 72, 74, 77–8, 81, 82
technologies, household 193–5, 203
technology matrix 193, 194
terms-of-trade changes 237
terrorism
    homeland security 8, 29, 82, 91, 108,
        116, 126, 159, 258
    insurance 24, 29
    optimal protection 2–3, 71–82
    three-ports study 147, 149–50
    utility lifelines 187–8, 190, 193
    see also September 11 (2001)
Thames Barrier 1, 2, 4
    analysis 126–9
    closures (incidence) 121, 122
    construction costs 120–22
    Great Flood (1953) 123–5
    implications for Katrina 131–2
    legislation 120
    location 129
    monitoring and operation 122
    proposals for future protection 129–31
    role of Sir Hermann Bondi 125–6

Thames Gateway 131, 132
three-ports study 147, 149–50
Tiebout model 219–20
tornado damage 25
total attacker utility 73, 74
tourism 4–5, 134–45, 147, 236, 289
trade theory 238
traffic (evacuation lessons) 67–8
tragedy of the commons 238
training 115, 116
tipping/sorting models 217
transport
    air 81, 108, 112, 131, 134–9, 142–5
    costs 152, 189
    evacuation 93, 95, 105, 106
    lessons learned 67–8
    modes 109, 111–12, 115, 204
    needs 93, 95, 105, 107, 108–9, 204
    private 193, 204
    roads 4, 106, 107, 128
    routes 3, 93, 106, 107, 115, 116
    sea 125, 127–8, 147, 149–50, 159
Transport for London 127
trip generation/distribution 3, 93
Tropical Storm Alison 12
tsunami 234
    Indian Ocean (late 2004) 208, 231,
        249
Tulane University 94, 97–8, 282
tunnel vision 243–5
Twin Towers (responses) 91
two-target game 79–81
Typhoon Bart 12
Typhoon Mireille 12
Typhoon Songda 12
typhoons 11, 12

uncertainty 95, 96, 203, 249
    analysis 50
    insurance and 25–6
under-seepage flows 63
underwriting 13, 14
unemployment 89, 265, 268, 280,
    282
United Airlines Flight 93 115
United Nations 233
University of North Carolina 94, 97,
    99–100
Urban Planning Committee 288
US Air National Guard 112

US Army Corps of Engineers 2, 34–6, 54–5, 60, 85, 132, 231, 240, 243, 248, 254
US Bureau of the Census 44, 113
US Bureau of Economic Analysis 44, 175
US Bureau of Labor Statistics 44, 283
US Coast Guard 91, 112
US Congress 126
US Department of Commerce 283
US Department of Homeland Security 8, 82, 91, 108, 116, 126, 159, 258
US Department of Housing and Urban Development 258, 260, 262–4, 275, 287
US Environmental Protection Agency 129
US Geological Survey (USCG) 39
US Government Accountability Office 21, 71
US Green Building Council 275–6
US Interagency Advisory Committee on Water Data 40
US Navy 112
US Postal Service 282
US Property and Casualty insurance industry combined ratio 13–14
US Treasury 17
USC sectors 150, 152–3, 158, 178–86
utility lifeline services (disruption) 5, 187–205

value-added 188
victims
  deaths 9, 12, 53, 60, 86, 89, 107, 115, 123, 208
  institutional disasters 88–90
  missing persons 12, 115
  rescue of residents/visitors 110–12
Vietnam War 208–9
vigilance, resilience and 204
visitors, rescue of 110–12

volcanic eruptions 234
  Eastern Caribbean 243, 245–7
voucher schemes 21, 275, 287
vulnerability 6, 230, 231, 232, 236, 239, 284

Wal-Mart 287
Wallace v. Blanco case 283
wash-out 62
water system 67
  supply 188, 196–8, 199, 201–3, 204–5
Waterborne Commerce of the US (WCUS) 152
Waverly Committee 128
welfare measures (of impacts) 189
Wharton Risk Center 10
Wharton School 23
wind
  damage 213–14, 253–4, 279
  insurance 15–16, 24–5
  speed/strength 40–41, 55, 56, 253
  -water controversy 24, 25
  see also hurricanes; typhoons
Winterstorm Daria 12
Winterstorm Lothar 12
Winterstorm Vivian 12
Woolwich Reach 129
work structure (New Orleans) 89
workmanship 37, 38, 39, 42, 43
World Bank 235, 244
World Heritage Site (Greenwich) 129
World Institute for Strategic Economic Research (WISERTrade data) 152–3
World Trade Center
  attack (1993) 81
  September 11 (2001) 2–3, 5, 8, 10, 11, 12, 71–82, 91, 115, 144
worldwide insured losses 9–10
worst-case scenarios 3, 84–92, 115

zip code areas 99, 100, 101, 282–3